SYMMETRY IN CHAOS

p112 . may apply elsewhere

SYMMETRY IN CHAOS

A Search for Pattern in Mathematics, Art and Nature

MICHAEL FIELD AND MARTIN GOLUBITSKY

Oxford New York Tokyo
OXFORD UNIVERSITY PRESS
1992

Oxford University Press, Walton Street, Oxford OX2 6DP

Oxford New York Toronto
Delhi Bombay Calcutta Madras Karachi
Kuala Lumpur Singapore Hong Kong Tokyo
Nairobi Dar es Salaam Cape Town
Melbourne Aukland Madrid

and associated companies in
Berlin Ibadan

Oxford is a trade mark of Oxford University Press
Published in the United States
by Oxford University Press Inc., New York

A catalogue record for this book is available from the British Library

Library of Congress Cataloging in Publication Data

Field, Michael.
Symmetry in Chaos/Michael Field and Martin Golubitsky.
1. Symmetry. 2. Chaotic behavior in systems.
I. Golubitsky, Martin, 1945– . II. Title.
Q172.5.S95F54 1992 003'.7—dc20 92-14023CIP

Designed and typeset by DTP by
Pete Russell, Faringdon, Oxfordshire.
Computer images generated from data supplied by the authors, and translated for film output by Focal Image, London.

Printed in Hong Kong

To April and Barbara

Preface

IN our mathematics research, we study how symmetry and dynamics coexist. This study has led to the pictures of symmetric chaos that we present throughout this book. Indeed, we have two purposes in writing this book: to present these pictures and to present the ideas of symmetry and chaos—as they are used by mathematicians—that are needed to understand how these pictures are formed. As you will see, the images of symmetric chaos are simultaneously complex and familiar; the complexity stems from the chaotic dynamics by which the pictures are produced, while the familiarity is due to the symmetry.

Although symmetry has long played a key role in mathematics and indeed in virtually all models of the universe, the study of chaotic dynamics in mathematics and its use in modelling physical phenomena is a more recent endeavor. It is worth noting that both words *symmetry* and *chaos* have standard meanings in the English language as well as technical definitions in mathematics. There are clear similarities between the everyday usage and the technical definitions for each of these words—but the similarity is rather more for symmetry and rather less for chaos. In both everyday usage and mathematics, symmetry has the sense of repetition. For example, symmetry gives unity to designs from the rose windows of great cathedrals to the wallpaper in your own home by repeating one design a large number of times.

Chaos, on the other hand, means `without form'—the great void. Viewed in this light, it is difficult to see how chaos can be the subject of scientific inquiry—which is based on finding form and regularity in the physical world. In recent years, the term chaos has (perhaps unfortunately) been adopted by mathematicians and scientists to describe situations which exhibit the twin properties of complexity and unpredictability. Archetypal examples are the weather and the stock market—although complex and unpredictable, these examples are far from being without form or structure.

One of our goals for this book is to present the pictures of symmetric chaos—in part because we find them beautiful and in part because we have enjoyed showing them to our friends and think that may appeal to others. But we also want to present the ideas that are needed to produce these pictures. For although the methods by which these computer-generated images are obtained are relatively simple, it is difficult to conceive how they might have been discovered without an appreciation of the underlying mathematics on which they are based.

In the first two chapters we discuss in general how the pictures are produced and how they are related to the mathematical ideas of symmetry and chaos. The third chapter has the role of an intermezzo. As we noted previously, the images of symmetric chaos often seem quite familiar. So in Chapter 3 we have paired off a number of pictures from nature (diatoms, shellfish, flowers, etc.) and a number of decorative designs (from rose windows to corporate logos to ceramic tiles) with designs produced on the computer using symmetric chaos methods. The fourth chapter is devoted to a more detailed discussion of the simplest forms of chaotic dynamics.

As you will see, we use three mathematically different methods for computing our images. Indeed these images which we call *symmetric icons*, *quilts*, and *symmetric fractals* are quite different in character. The methods for producing the icons, quilts, and symmetric fractals are explained in detail in the last three chapters. Since many of the readers of this book may be familiar with fractal art—say as appears in the books *The Beauty of Fractals* by Heinz-Otto Peitgen and Peter Richter and *Fractals Everywhere* by Michael Barnsley—it is worth noting that the images we present have a different character from those found in fractal art. While fractal pictures have the sense of *avant garde* abstract modernism or surrealism, ours typically have the feel of classical designs.

In the first appendix we present the exact parameter values that we have used to produce the pictures of symmetric chaos found in this volume. In Appendix B we give detailed computer programs (written in QuickBasic) that will enable the reader to experiment on a home computer with the formulas for symmetric chaos presented in Chapters 5-7. The actual derivation of these formulas is found in the last two appendices—one for the icons and one for the quilts. These sections contain more technical mathematics than the previous chapters.

Though this book is concerned with symmetry and chaos and their relationship with pattern formation and geometric design and art, we have not attempted here to describe in depth the many possible threads that lead from this work, in part, because these issues have been discussed elsewhere. Indeed, many authors have written about symmetry and art, but perhaps none more elegantly than Hermann Weyl in his classic book *Symmetry*. There have also been a number of books on chaos—our favorites being Ian Stewart's *Does God Play Dice?* and James Gleick's *Chaos: Making a New Science*. Finally, the mathematical and scientific discipline of pattern formation, which underlies our own work, is discussed in *Fearful Symmetry: Is God a Geometer?*

There are a number of individuals whose help we wish to acknowledge. This help has taken a variety of forms from deriving the basic theory on which our pictures of symmetric chaos are based, to helping with the computer programming needed to produce high resolution color graphics on workstations, to making helpful suggestions on how better to use color in our pictures. We thank Peter Ashwin, Pascal Chossat, Robert Cottingham, Michael Dellnitz, April Field, Nathan Field, Michael Flanagan, Elizabeth Golubitsky, Phil Holmes, Barbara Keyfitz, Greg King, Martin Krupa, Ian Melbourne, Ralph Metcalfe, Jim Richardson, Harry Swinney, Hans True, and, in particular, Ian Stewart. We also thank Wendy Aldwyn, who has produced the hand-drawn art work including several original drawings and has helped us with the coloring of several of the computer drawn pictures.

Layout, mathematical notations, graphics and color reproduction have made this book unusually complex to produce. Special thanks are due to Oxford University Press for their comprehensive assistance and helpful suggestions at every stage of the production process. The computer generated pictures were produced directly from computer files by Kaveh Bazargan of Focal Image, Ltd. using state of the art processes. Finally, we acknowledge the institutional assistance of the Mathematical Sciences Institute, Cornell University; the Department of Mathematics, University of Houston; and the Department of Pure Mathematics, University of Sydney, for providing the intellectual and computer environments needed to produce these pictures.

Sydney and Houston M.F.
February 1992 M.G.

Contents

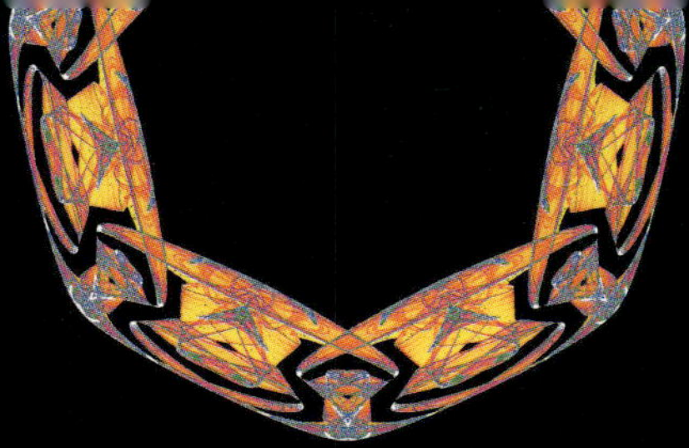

Chapter one

INTRODUCTION TO SYMMETRY AND CHAOS

THE pictures shown in Figures 1.1 and 1.2 were drawn by a computer. When we respond to a work of art, be it a painting, mosaic, or sculpture, we often share with the artist various sentiments as to what combinations of form and shape are appealing or otherwise dramatically effective. We see in these pictures that the computer is able to create shapes and designs that can mimic patterns observed in nature or created by man. Yet, these pictures were not created by an artist. Instead, artistic sensibilities and aesthetic judgements have, to a large extent, been replaced by precise mathematical formulas. Intriguing though the question is, we shall not venture any guesses as to why this mimicking takes place. Instead, our goal will be to describe how these pictures are formed. Along the way, we shall make contact with some of the most fascinating ideas of twentieth century mathematics and science.

Our pictures are created by merging symmetry and chaos. At first sight, this seems paradoxical: a merging of order and disorder or yin and yang. To make any sense of this description, we must start by examining the ideas of symmetry and chaos.

Symmetry

We begin with a dictionary definition of symmetry (quoted from *The American Heritage Dictionary*):

symmetry *n.* Exact correspondence of form and constituent configuration on opposite sides of a dividing line or plane or about a center or an axis.

A first reading may suggest that dictionaries are just masters of obfuscation! Yet, symmetry is a very basic concept, like that of

Figure 1.1 (opposite) *Halloween.*

Figure 1.2 *Mayan bracelet.*

oneness, and it is notorious how difficult such ideas can be to explain. Naively, one tends to think of a geometric object as being symmetric if it is possible to cut it along a line or plane and get two identical pieces. However, this is too imprecise to serve as a working definition, as can be seen by looking at the mosaic pattern displayed in Figure 1.3, which is certainly symmetric but cannot be cut along any line to obtain equal pieces. Perhaps the simplest way to explain symmetry is to follow the

Figure 1.3 *Mosaic pattern formed from staplers with no lines of symmetry.*

operational approach used by mathematicians: a symmetry is a *motion*. That is, suppose you have an object and pick it up, move it around and set it down. If it is impossible to distinguish between the object in its original and final positions, we say that it has a symmetry. Thinking of symmetry in this way, we see that every object has at least one *trivial* symmetry obtained by picking it up and putting it back in its original position. Even though, at first sight, it may seem bizarre to say that everything has at least one symmetry, this convention is actually quite useful. Indeed, it parallels an important discovery of Indian mathematics: the number zero.

Next let us look at an object which has symmetries other than the trivial symmetry. We shall search for the symmetries of the three-pointed star shown in Figure 1.4. We have labelled the points of the star A, B, C to aid our discussion. If we rotate the star clockwise through one-third of a turn about its center O, we move it to a position that is geometrically indistinguishable from its original position. However, point A is moved to where point C was, point C to where point B was, and point B to where point A was (see Figure 1.5(a)). This motion is a *nontrivial* symmetry of the star. We may repeat the process and further rotate the star clockwise through another one-third of a turn (that is, the star will have been rotated clockwise two-thirds of a turn from its original position). We see that point A has been moved to B, point B to C and point C to A (see Figure 1.5(b)). If we again rotate the star by a third of a turn, it will return to its original position. Thus far we have found three symmetries: leave fixed, rotate one-third of a turn, rotate two-thirds of a turn. However, these are not the only symmetries of the star. Referring

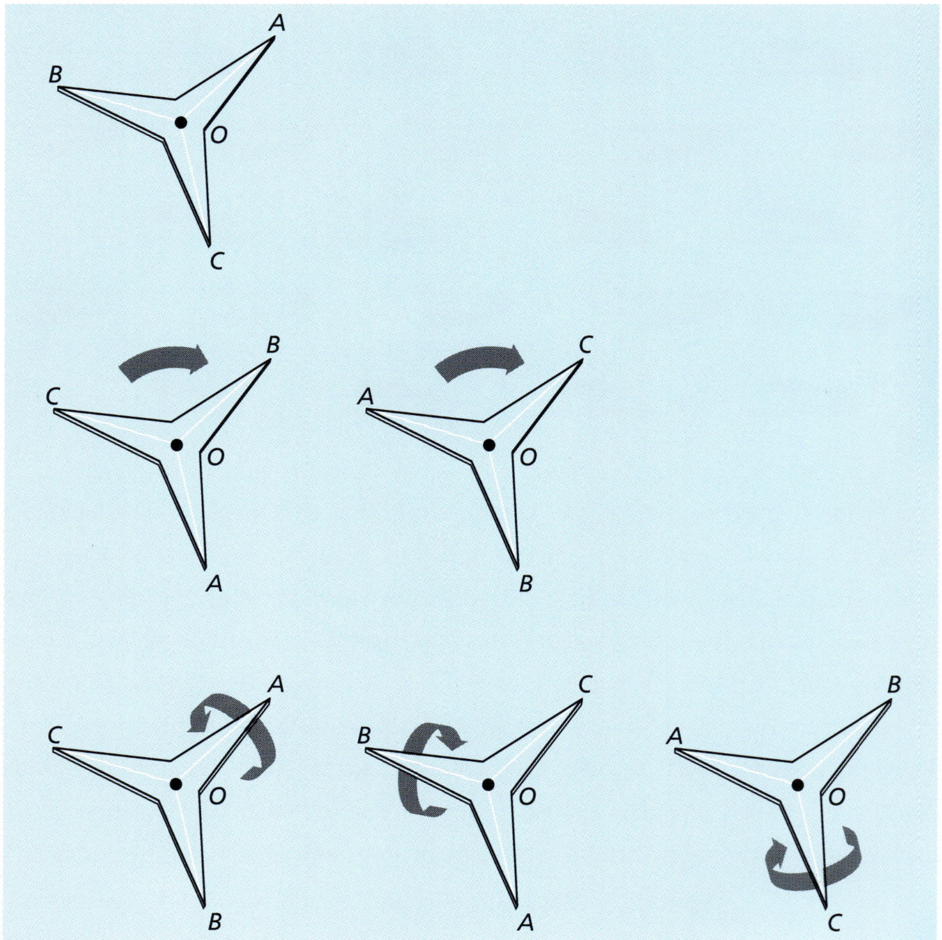

Figure 1.4 *A three-pointed star.*

Figure 1.5 *Three-pointed star rotated by (a) 120° (b) 240°.*

Figure 1.6 *Three-pointed star reflected across (a)* AO-*axis; (b)* BO-*axis; (c)* CO-*axis.*

to Figure 1.4, observe that we can *reflect* the star through the line *AO*. Otherwise said, hold the star along the line or *axis AO* and make one half-turn in space so as to interchange *B* and *C*, keeping *A* fixed (see Figure 1.6(a)). That is, we flip the star upside-down, keeping the line *AO* fixed. Clearly, there are two other reflection symmetries of the star obtained by reflecting in the lines *BO* and *CO* (Figures 1.6(b,c)). It turns out that the six symmetries that we have found are all the symmetries of the star.

Of course, the six symmetries of the star that we have described are precisely the symmetries of an equilateral triangle and are common to many figures. In particular, the color plate Figure 1.7 has exactly the same symmetries as the star. On the other hand, the next color plate Figure 1.8 does not. While it has the rotational symmetries of the star, it does not have the reflectional symmetries. However we reflect the figure (through any line), we always get a different figure. It may be

Figure 1.7 (opposite) *Wild Chaos.*

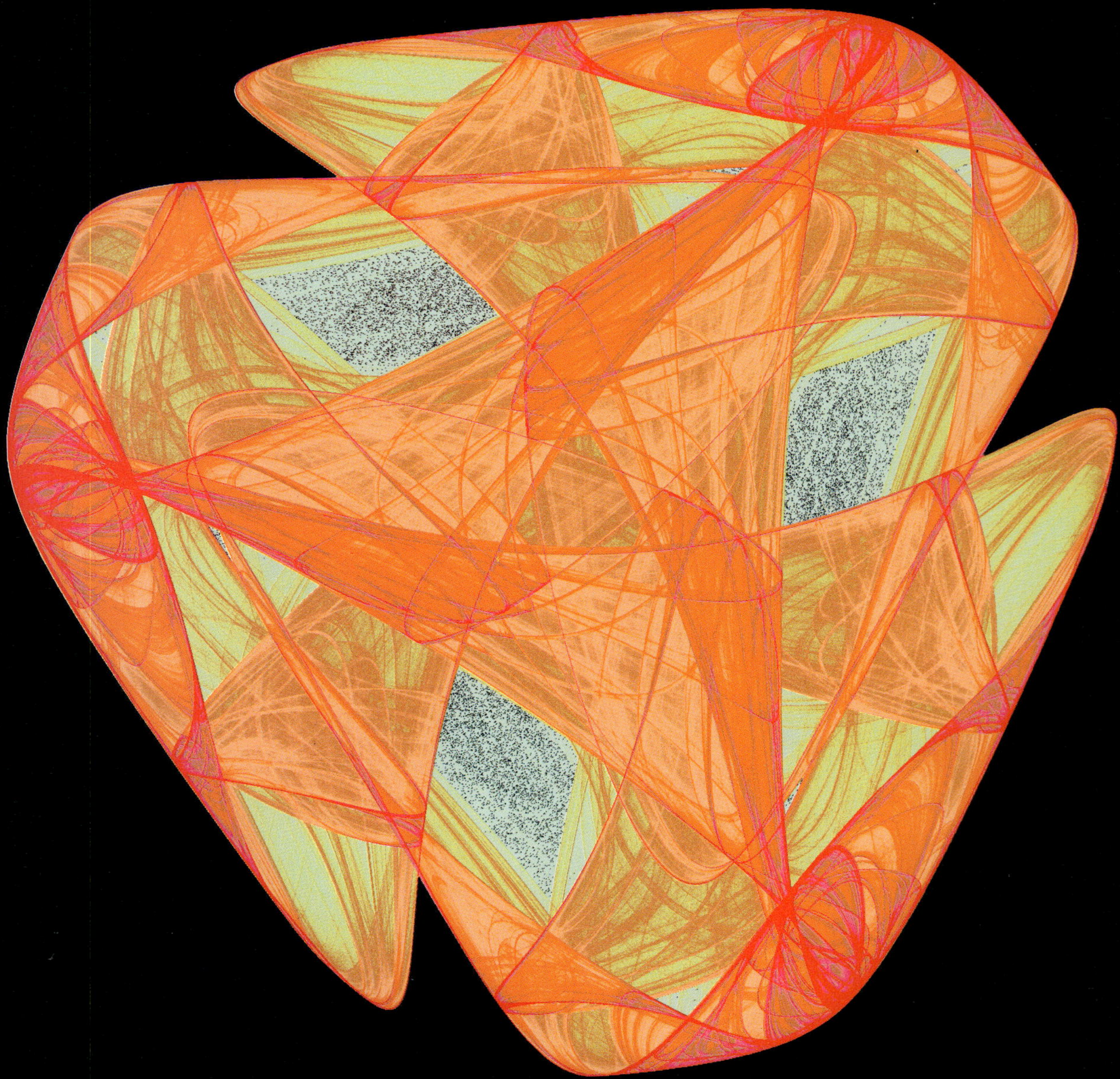

Figure 1.8 *Clam Triple. Chaos without reflectional symmetries.*

helpful to note that the figure appears to come with a *counterclockwise* twist. After being reflected, the figure has a *clockwise* twist.

Subsequently, we shall look more carefully at figures with different symmetries (for example, the symmetry of the square). Symmetries of one type or another figure prominently in many artefacts created by humans. Even the ubiquitous road sign invariably has some symmetry. At first glance, symmetry is not immediately apparent in nature. For example, geographical features such as the landscape, mountain ranges and oceans are far from symmetric. Living organisms often have approximate symmetry—usually they resemble their mirror images —but the correspondence is rarely exact. One of the most obvious, and beautiful, examples of symmetry in nature is the fantastic symmetry of snowflakes (see Figure 1.9). It is perhaps not so surprising that symmetry can be found in the physical world, because symmetries are woven into the physical laws that govern the universe. However, in the acting out of those laws, much of this symmetry becomes invisible to us. In some way, our response and feeling for symmetry is perhaps a reflection of the underlying symmetry of these physical laws.

Figure 1.9 A Snowflake with hexagonal symmetry.

Chaos

Next we describe what, at first sight, is the antithesis of symmetry: chaos. Again, we begin with a dictionary definition:

chaos *n.* Any condition or place of total disorder or confusion.

Whatever else, it would appear that chaos is featureless, without form or structure. Over the past decade or so, the term *chaos* has been used increasingly in science and mathematics. It has been suggested that a wide range of natural phenomena, from the orbit of Pluto to the weather or reversals of the earth's magnetic field, are chaotic. The use of the word *chaos* in this context is a little misleading. The very word *chaos* suggests a lack of form or feature; however, we think of scientific investigation as being restricted to those phenomena that do have definable structure or features. Crudely put, if something is in a state of total disorder and lacks structure or form, there is not much we can say about it.

In this book we shall use the term *chaos* in the sense that it is used in contemporary science. Roughly speaking, we regard chaos as being characterized by unpredictability and complexity. To proceed further and give more precision to our description we need to go back to the time of Isaac Newton and the birth of the modern viewpoint on *dynamics* or the evolution of a system in time.

Determinism

Newton not only formulated his famous theory of gravitation, and three laws of motion, but also invented the mathematics that enabled him to develop the consequences of those laws. For example, he gave a precise mathematical model for the motion of the planets around the sun. Perhaps the most important feature of Newton's model was its ability to give exact predictions of the future orbits of the planets. In other words, if we know the position and speed of the planets at any given time we can, in theory, predict the positions of the planets at any subsequent time. Thus, Newton's model for the solar system is *deterministic*: if we know the initial position and velocity of the planets, then their subsequent motion is uniquely determined.

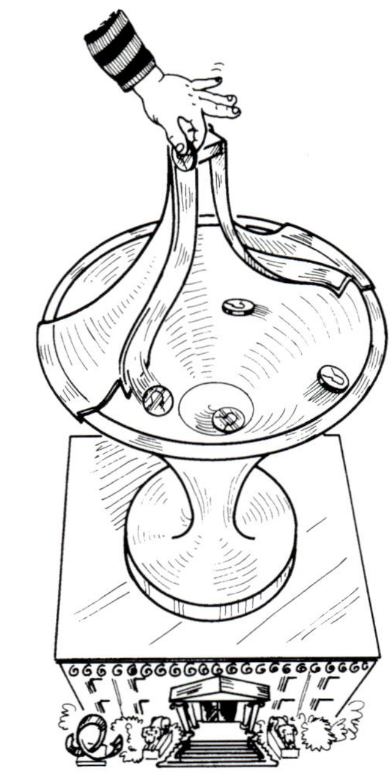

Figure 1.10 *A gravity well.*

This point is brought out more clearly if we look at a non-deterministic model. For example, it is a consequence of the rotation of the earth that the sun always rises in the east. Suppose, instead, that the matter of where the sun rises depended on the toss of a coin. That is, imagine that every night God tosses a fair coin. If the coin comes down heads, the sun rises in the east, if it comes down tails it rises in the west. This model for the rising of the sun is non-deterministic. Whether the sun rose in the east or west today has no bearing on whether it will rise in the east or west tomorrow.

A common expectation, based on determinism, is the belief that reasonable knowledge of current position leads to good predictions of future position. In other words, outcomes should be the same given roughly the same initial data. For example, the gravity well, pictured in Figure 1.10, is a toy that, in recent years, has appeared spontaneously in museums throughout America, with the purpose of making the act of giving more enjoyable. Typically, the gravity well is put into action when a parent gives a coin to a child who then hurls it along the ramp located at the top of the well. After launch the coin spins its way to the bottom, rotating ever more quickly, much to the delight of children and parents alike. At the bottom the coin falls, kerplunk, into a box. Different speeds of entry of the coin lead to slightly different trajectories, but ultimately to the same kerplunk. Indeed, the purpose of moving coins from your pocket to the museum coffers would not be well served were the final state not independent of the initial conditions.

The color plates shown in Figures 1.1 and 1.2 were generated by a deterministic model, in this case by a precise mathematical formula. However, in spite of this determinism, the model behaves in many ways in a non-deterministic or random manner. This is characteristic of chaotic systems: they are deterministic but behave as though they are not. How can determinism produce a seeming lack of determinism? The answer, in part, depends on the fact that in practice we can never know exactly our initial position or state. A tiny inaccuracy in our knowledge of current position can be magnified to produce completely inaccurate predictions of future position. Even if we do know our current position exactly, when we come to compute our final position—using a computer—small errors can and do creep into the arithmetic. In a chaotic system, these errors are rapidly magnified and force inaccurate predictions. The realization that to all intents and purposes a deterministic system can behave as though it were non-deterministic is relatively recent. In the nineteenth century the belief in a deterministic universe was so strong that the French mathematician and physicist Pierre Simon Laplace was able to say that

> such an intelligence would embrace in the same formula the movements of the greatest bodies of the universe and those of the lightest atom; for it, nothing would be uncertain and the future, as the past, would be present to its eyes.

The development of quantum theory at the beginning of the twentieth century showed that at the atomic level there is a basic lack of determinism in physical laws. More recently, it has become clear that, even if we ignore quantum effects, Laplace's optimistic view that the future can be predicted on the basis of enough information about the present is fundamentally flawed. The problem is a practical one. Even the tiniest error in our data can be magnified over time to produce an outcome far different from what we would have gotten using the true data.

Sensitive dependence

The sensitive dependence on initial conditions that is the hallmark of chaotic dynamics can have important implications for everyday life. Indeed, consider the weather. We know that temperatures rise and fall within broad limits dictated by the seasons and that rain and sunshine alternate. But, for a variety of reasons, we would like to have more precise knowledge. 'Will it rain on Thursday afternoon?', 'How much snow will fall on Sunday?', and 'How hot will it be on Tuesday?' are typical questions. The answers to these questions depend crucially on having exact knowledge of current weather conditions, which is literally impossible, given the complexity of such data.

Suppose that the weather system is chaotic, which many scientists believe to be the case. Then accurate and detailed long-term predictions of the weather are also impossible. We are faced with the possibility that sensitive dependence on the initial data makes accurate predictions of the weather five or six days in advance a practical impossibility.

One indisputable feature of the weather is its complexity and this is a characteristic feature of chaotic dynamics. Complexity is certainly a feature of the color plates displayed throughout this book. On examination, we see that each picture has a most complicated and intricate structure. This complexity and structure tell us that the pictures contain *information*. The problem, just as with the weather, is how to use and interpret this information. Contrary to common belief, uncertainty is often *proportional* to the amount of available data rather than in inverse proportion! Moreover, our response to complicated phenomena often compounds the difficulty.

Information concerning the economy is a good example. Every day we are bombarded by the media with an abundance of data—

data about the stock market, unemployment rates (local and national), the gross national product, foreign trade, inflation, taxes, worker productivity, banking practices, and so on. In fact, for most of us, the effect of this mass of data is to make us even less certain about what is actually happening to the economy. In an attempt to control the mountain of data, we, as well as our leaders, tend to look for simple indicators that will predict the direction of the economy. In short, our solution to this uncertainty is often to base all argument and prediction on just one or two features ('the money supply' or 'inflation'). While this approach cannot be said to be scientifically valid, it has the merit of being comforting and very flexible.

Again it is appropriate to stress that we draw a sharp distinction between the dictionary definition of chaos and the sense in which we use the term here. The dictionary definition of chaos implies that there is no information within chaos: it has neither form nor structure. For us, chaos may be complex and appear to be non-deterministic, but hidden within it is a wealth of information.

Rules

Before we can come to grips with explaining why our pictures represent chaotic dynamics, we need to spend some time talking about how the pictures are created. Computers really are obstinate beasts. There is an old joke about a professor complaining to a colleague: 'My student is incredibly stupid; I've taught him everything I know and he still doesn't know a thing.' Computers are exactly like this. Anyone who has worked with one knows that the 'black box' is remarkably dumb—it does precisely what it is told, nothing more, nothing less. To create

Figure 1.11 *Pixels on a 10 × 10 grid.*

pictures like the ones here, we have to tell the computer precisely what to do, correcting our own mistakes along the way.

What shall we tell the computer to do? Basically, the computer must be told which lights on the screen it should turn on and in which color. Each light on the screen is called a *pixel*, and a modern high resolution monitor has about one million pixels each of which may be thought of as a square about 0.4 mm across (or about 65 to the inch). Because the pixels are so small, it is often convenient to think of them as representing points on the screen. In Figure 1.11, we show how the pixels are arranged on a monitor screen with 100 pixels in a 10×10 grid.

The way we tell the computer to make pictures is by using mathematical formulas. These formulas determine the pictures that are drawn. A *symmetric* formula can produce a symmetric picture. Abstractly, formulas provide a shorthand for the rules of arithmetic and these rules of arithmetic provide us with a mechanistic way of choosing points in the plane or pixels on a computer screen.

The word *rule* has several meanings, the most familiar being restriction. As an example, consider a basic rule in many homes: no feet on the furniture. (In this sense, rules are usually unfair!) However, the type of rule we have in mind is a set of instructions. For example, our GO TO BED rule is: go upstairs, have a bath, brush teeth, do not read, turn off the lights, get in bed, and go to sleep. This rule is quite complicated. To learn it takes at least twelve years of constant reinforcement.

The GO TO BED rule is, in one respect, quite simple. Once invoked, there is only one allowed outcome (sleep). Arithmetic rules are more complicated: answers depend on inputs. A calculator with a square root button provides a good example of an arithmetic rule. The rule is: enter a number and push the square root button. Should we enter 4, out will come 2; if we enter 2, out will come 1.414 213 5 (or something close, depending on the calculator). Of course, we can apply the square root rule many times by just repeatedly pressing the square root button. When we do this, we are simply taking the square root of the result from the previous calculation. Indeed, this repetition will eventually lead to the calculator displaying the number 1 (unless you started with a number less than or equal to zero). From now on we shall use the word *rule* only in the sense of one of these mathematical rules. In particular, a rule will always have an input (typically a number) and an output (typically a number).

A good example of an arithmetic rule is provided by the following tale from folklore. A wise man, using his wit, is able to save a (somewhat slow and unpopular) Indian king from devastation (you may choose from what and how). In return, the king offers him half his kingdom worth 500,000,000 rupees. The wise (and shrewd) man modestly asks the king: 'Oh, honorable king, I am but a poor man. Perhaps, instead, you could just fill the squares on a chess board for me, starting with a grain of rice on the first square, two grains on the second, four grains on the third, and so on.'

As king, do you grant the wise man his wish, put him in charge of the national debt, or chop off his head? The difficulty lies in the 'and so on'. By the tenth square, the king would be giving the wise man a modest 2047 grains of rice. However, by the twentieth square, he would be giving 2,097,151 grains of rice. By the last square, there would be enough rice to cover all of India fifty feet deep. Put another way, if the grains of rice were laid end to end they would stretch from the earth to the sun and back, thence to the nearest star Alpha Centuri and back, and then again to Alpha Centuri and back, with enough rice left over to provide biryanis for the king and his courtiers for the rest of their lives.

More formally, the wise man asks the king to start with a grain of rice and double that amount 63 times. As long as you start with a nonzero number, the doubling rule makes numbers grow larger and larger. As the wise man shrewdly observed, the numbers grow unimaginably large quite quickly.

Pixel rules

If we intentionally confuse pixels on the screen with points (in the plane), then the rules that make our pictures are similar to the arithmetic rules we have described above. However, unlike the doubling rule, we do not want our rules on repeated application to grow without bound (otherwise, points would soon leave the computer screen). We think of a pixel rule as a rule which has pixels as input and pixels as output. The pixel rule may depend on some complicated mathematical formula but for the moment we wish to keep the arithmetic hidden. To make a black and white picture, we assume that the screen is black. We choose one pixel and turn it on—the corresponding point on the screen will then be white. Then we invoke our pixel rule, beginning with the first pixel as input, and obtain a new pixel which we turn on. Finally we repeat this rule over and over again until we decide to stop. The whole process is called *iteration*. In this scheme there is no reason why one pixel cannot be visited more than once.

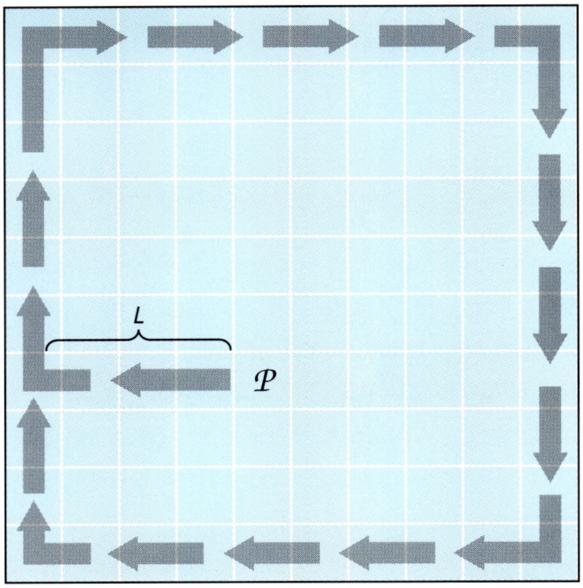

Figure 1.12 *Dynamics on the pixel grid.*

As an example of a very simple pixel rule, choose one pixel from the screen, say the top left. We define a pixel rule by requiring that whatever pixel we choose from the screen as input, we always get the top left pixel as output. However many times we apply the rule, we never see more than two pixels lit on the screen: the initial pixel and the top left pixel.

Next we look at a slightly more complicated pixel rule. Following Figure 1.11, suppose that the monitor screen has 100 pixels arranged in a 10×10 grid. Choose a pixel $\mathcal{P}$ from the screen and the direction left. The pixel rule has two parts: if you can, move one pixel in the direction you are going; and if you cannot, turn right one-quarter of a turn. The picture that will result from this pixel rule is simple to describe. There is an initial segment moving left from the initial point $\mathcal{P}$ to the boundary of the grid followed by a never ending circumnavigation of the boundary in the clockwise direction (see Figure 1.12).

Even though the rules we have described here are rather simple, there are one or two interesting features that we want to single out for special mention.

First of all, note that the first part of the pixel sequence is different from its long-term behavior. In particular, the pixels on the initial line segment, labelled L in Figure 1.12, are never revisited. We say that this part of the pixel sequence represents the initial behavior. We often use the term *transient* to describe the initial behavior. The transient behavior is seen at the beginning but not in the long term. The part of the pixel sequence beginning at the boundary represents the *long-term behavior*.

A second important observation about this example is that the long-term behavior repeats *ad infinitum*. We refer to this characteristic as *periodicity*. Since there are 36 pixels on the perimeter, this pixel rule repeats itself every 36 iterates (ignoring the initial transient).

Indeed if we apply any pixel rule enough times, then eventually at least one pixel will be revisited. To see why this is so, suppose that there are 100 pixels on the screen (any large number will do equally well). After 100 iterates we have 'lit' 101 pixels, so at least one pixel must have been 'lit' twice. (This argument is an example of the *pigeonhole* principle: if there are 101 letters to be put in 100 pigeonholes, at least one pigeonhole must contain at least two letters.) It follows that if the rule that we used to create our pictures were actually a pixel rule, then, after an initial transient, we would have to find periodic behavior. In general, our picture rules do not lead to this simple kind of periodic behavior, and color can be used to understand this point.

Coloring by number

We now say more about how we color our figures. The basic idea is quite simple. Start with a mathematical formula generating a picture such as Figure 1.13. Choose an initial point and apply the rule a large number of times, typically between 20,000,000 and 100,000,000. Ignore the transient part of the pixel sequence that is produced. (In practice, we only count pixel hits after the first 1000 applications of the rule.) Record the number of times each pixel is hit, and color the pixel according to the value of that number. This process is no more than

Figure 1.13 Emperor's Cloak. Pentagonal symmetry and a color bar.

coloring by number. The actual colors are chosen according to which colors bring out best the underlying structure. In Figure 1.13, we show the result of coloring a figure with fivefold symmetry after 30,000,000 iterations. Since there are less than 1,000,000 pixels, it follows by the pigeonhole principle that some pixels must have been hit more than once. In practice, many pixels are hit more than once, and in the color band in Figure 1.13, we show the colors assigned to pixels, depending on the number of times they have been hit. As always, we leave the pixel black if it has not been hit. We color white, shading to yellow if the pixel has been hit between 1 and 15 times; yellow shading through to red if the pixel has been hit between 16 and 135 times, and finally through to navy blue if the pixel has been hit more than 585 times. In this particular picture, some pixels are hit 7637 times.

Thus far, we have confused pixels and points on the screen and regarded our mathematical formula as a pixel rule. However, when we make a large number of applications of our rule, we really have to distinguish the underlying arithmetical rule from a pixel rule. To see why this is so, recall that a pixel rule starts with a transient and then behaves periodically. A consequence is that the only sensible choice of coloring for pixels chosen using a pixel rule would be one color for the transient pixels (those visited only once) and another color for the pixels that are visited periodically. If we look at the colorings of Figure 1.13, we see that the picture represents a process that is far from periodic.

Arithmetic rules

To understand how we can create pictures with such intricate colorings, we need to discuss how our hidden arithmetic rule can be related to a pixel rule. In an arithmetic rule there is a formula that tells us how to move points in the plane to points in the plane. Suppose we choose an initial point to which we will apply an arithmetic rule. Since a pixel is actually a small square region on the computer screen, this point lies inside one of the pixels. (There is a small difficulty concerning what happens when the point is on the common boundary of two pixels, but we ignore this issue.) Applying the arithmetic rule, we get a new point on the screen which lies inside a single pixel. We turn on this pixel. If we

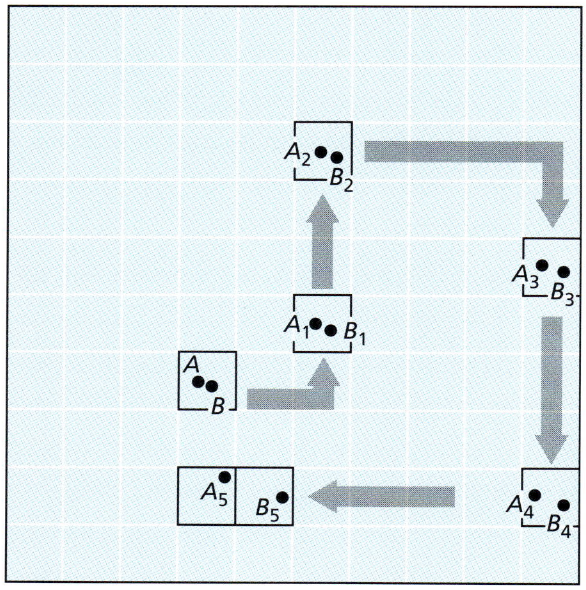

Figure 1.14 *The divergence of points within a pixel on iteration.*

repeat this process many times, eventually there will be a pixel which is visited twice. As we explained above, if we were working with a pixel rule, it would follow that subsequent behavior would be periodic. However, in our visualization of the arithmetic rule, it is important to understand that, even though a single pixel may be visited many times during the iteration process, no point inside that pixel need be hit more than once during this process.

Even though we do not want our arithmetic rule to behave like the doubling rule—that is, points should not become unboundedly large and leave the screen—it turns out, perhaps surprisingly, that the rules we use to generate our pictures have much in common with the doubling rule. In Figure 1.14, we show two points A and B lying inside a pixel. For simplicity, we suppose the pixel has edge length equal to 0.4 mm. Suppose the distance between A and B is 0.02 mm. We now apply our arithmetic rule to find new points A_1 and B_1 corresponding to A and B respectively. Since A and B are close together, we would expect that A_1 and B_1 are likely to lie within the same pixel. However, the distance between A_1 and B_1 may be increased (see Figure 1.14). For the arithmetic rules used to produce our pictures, distances are often increased by at least some multiple. For example, the distance between A_1 and B_1 might be at least twice the distance between A and B. Now this property of doubling distances holds true provided that the distance between points is not too large (say, not more than one-tenth of the diameter of the screen). If we apply our arithmetic rule to A and B five times, and let A_5 and B_5 be the resulting points, then the distance between A_5 and B_5 would be at least 32 ($=2^5$) times the distance between A and B. A quick computation shows that this distance is at least 0.64 mm and so A_5 and B_5 must now lie in different pixels.

What this argument shows is that if our arithmetic rule tends to increase distances between points which are close together, then the eventual outcome of even relatively few iterations will depend very sensitively on the initial points. However, it is a remarkable fact that if we perform a very large number of iterations and color our pixels according to the number of times points meet each pixel, then the resulting picture is usually *insensitive* to the initial point chosen. To understand why this might be so we need to discuss two topics: strange attractors and statistics.

Strange attractors

It is indeed difficult to give a simple explanation of how symmetry and dynamics merge to generate both chaos and structure at the same time. Mathematicians and physicists are still trying to understand the methods whereby chaotic dynamics produces repeatable and observable patterns. Although chaotic dynamics has only become well known in the last decade, even this limited familiarity has led to some understanding, and much of this understanding is based on pictures.

In part, however, the situation is a little like old Uncle Jake who is a bit eccentric. You're not really surprised by what Uncle Jake does, but it's still difficult to understand why he does what he does. But by calling him eccentric, you feel comfortable with his actions. We often name issues or situations that are too complex to understand, and then feel at ease when these names are repeated. Well, mathematicians are sometimes as guilty as others in naming away their lack of understanding. In this case, we call those geometrically complicated objects that appear by iteration *strange attractors*. The word strange is used simply because they were unexpected! Why attractors? Well, that we can explain.

In a way the behavior we see with attractors is the opposite of the unboundedness in the wise man's doubling rule. Recall that the doubling rule takes a number and doubles it. If we start with the number zero, the result is just zero again. We call zero a *fixed point* for the rule: it is fixed, or unchanged, when we apply the rule. However, if we start with a very, very small number (for example, 0.000 000 01) and apply the doubling rule over and over again (say thirty-two times) then we arrive at a relatively large number (about 43). Thus, even though the number zero is fixed, nearby numbers move *away* from zero when we repeatedly apply the doubling rule. This process of moving away from zero is slow at first but eventually ends up proceeding quite briskly. We might call the number zero a *repelling* fixed point. As we apply the rule, numbers move away from zero.

An attractor captures the opposite behavior. As we apply a rule, we move closer, or are attracted, to the attractor. For example, we may consider the halving rule, which takes a number and halves it. Since half of zero is still zero, we see that zero is a fixed point for the halving rule. However, unlike the doubling rule, all numbers move closer to zero under repeated applications of the halving rule. In this case we call zero an *attracting* fixed point: under repeated application of the halving rule, nearby numbers are attracted to zero.

The main feature of this attracting fixed point is that whatever initial point you choose, iteration will bring that initial point closer and closer to zero. The long term behavior consists only of the fixed point. In a similar way, more complicated geometric objects can be attracting. Indeed, when we make our pictures, we start with some initial point and iterate, throwing away what we believe to be the transients. How do we know that they are transients? Well, if we start with another initial point and throw away the transients, we end up with the *same* picture. The totality of points that form the picture is called the *attractor*.

We can now understand part of the reason why we get the same picture when we choose two different initial points to start the iteration process. Indeed, suppose that we are using a rule for the iteration that has an attractor. Provided that our choice of initial point is not too far away from the attractor, we find that the iterates approach the attractor and then appear to 'bounce around' chaotically on the attractor. The *order* in which we visit pixels that comprise the attractor may depend on the choice of initial point. However, the chaotic behavior of the pixels on the attractor leads to the iterates eventually visiting all the pixels on the attractor. Suppose we use a black and white picture to display the attractor and color pixels white when they have been hit. If we ignore the transient part of the iteration (where we are approaching the attractor), we eventually turn on exactly those pixels that make up the attractor, irrespective of the initial point.

This discussion gives some insight into why we get the same black and white pictures; but why do we get the same *color* pictures? To understand this issue we need to discuss some *statistical* properties of the attractors.

Statistics

Let us start with a review of the method used to color our pictures. We take a large number of iterates, ask how many times each pixel is hit during the iteration process, and then color by number. As we have noted previously, each pixel actually represents a small square in the plane. Suppose we perform 1,000,000 iterates and find that a particular pixel $\mathcal{P}$ is hit 500 times. This means that 0.2% of the points visited during the iteration process lie in this particular pixel. We may interpret 0.2% as a probability in the following way. Suppose we perform an experiment using our computer. We start the computer iterating at some initial point and then at some 'random' time we stop the iteration process and record the last point visited. Then that last point will lie inside the pixel $\mathcal{P}$ with probability 0.002. Suppose, for example, that we color amber all pixels that have been hit 500 times. The information that this coloring is really telling us is that if we were to perform the experiment just described, then it would be *equally likely* that the last point visited would lie in any one of the pixels that are colored amber.

One of the fundamental ideas surrounding chaotic dynamics is that strange attractors have precisely defined statistical properties. There really is a sense in which this probabilistic notion can be made precise. It is much too complicated an issue for us to delve into further here, but what we can say is that there is a very good *mathematical* reason for coloring our pictures the way we have. We are measuring *approximately* the probability that a given pixel will be hit during iteration and exhibiting this information through colors.

A consequence of this statistical aspect of strange attractors is that we can expect to obtain the same picture (with very high probability) independently of the initial point we choose (as long as it eventually converges to the attractor). We obtain identical pictures, not because the order in which the pixels are visited in the attractors is the same (it isn't), but because the probability that we will visit any given pixel in the attractor during iteration is independent of the choice of initial point.

Symmetry on average

At this point, it is worth exploring how the statistical nature of strange attractors affects the basic property of the pictures we have shown: their symmetry. As we have seen, one way of thinking of a strange attractor is as the result of averaging. Rather than looking at a large number of iterates, let us instead look at the pictures we obtain by spending a small amount of time in the iteration process, that is, by doing only a few iterates. In Figure 1.15, we show the result of performing a fairly small number of iterations using a rule with threefold symmetry. In Figure 1.15(a), we have performed 200 iterations. No structure is apparent and, at first glance, the distribution of points appears somewhat random. In Figures 1.15(b,c), we have recorded 600 and 1500 iterations respectively. We see that by the last picture definite symmetry has appeared, though the picture is not completely symmetric. In Figure 1.16, we show the result of 20,000,000 iterations. This figure now has definite structure and threefold symmetry. Finally, we perform 60,000,000 iterations and color the resulting figure (Figure 1.17). The use of color has brought out many fine details that were not apparent in Figure 1.16. Color also shows us which of the points on the attractor are most likely to be hit during the iteration process—and which are least likely to be hit.

Indeed, one of the reasons why chaos and strange attract-

(a)

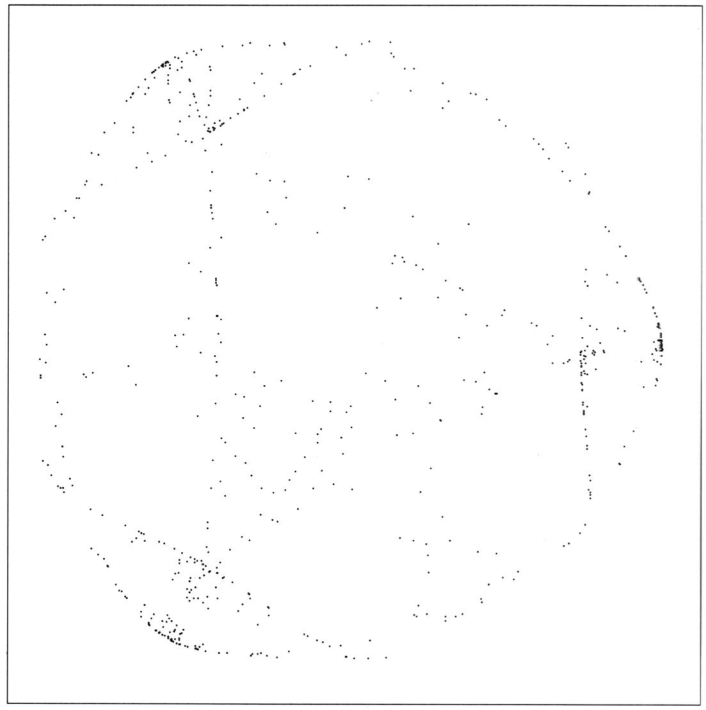

(b)

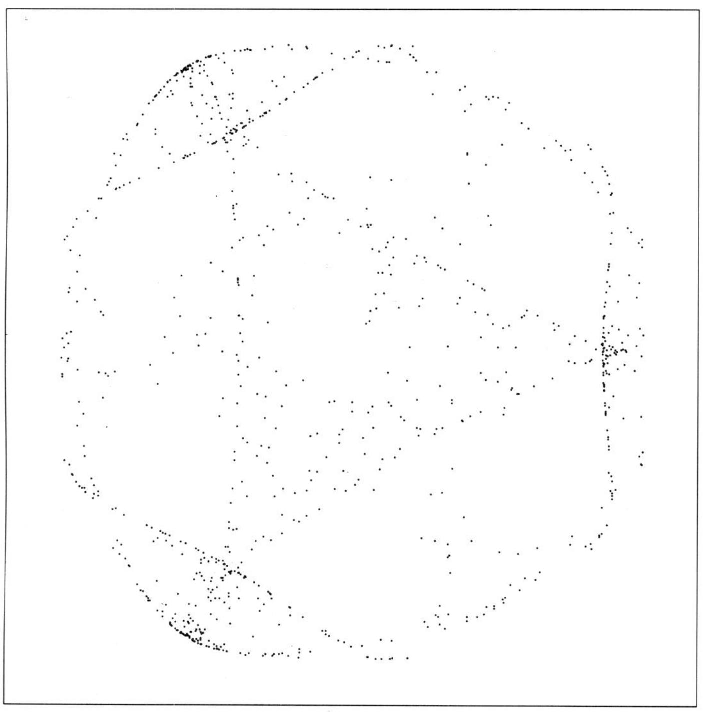

(c)

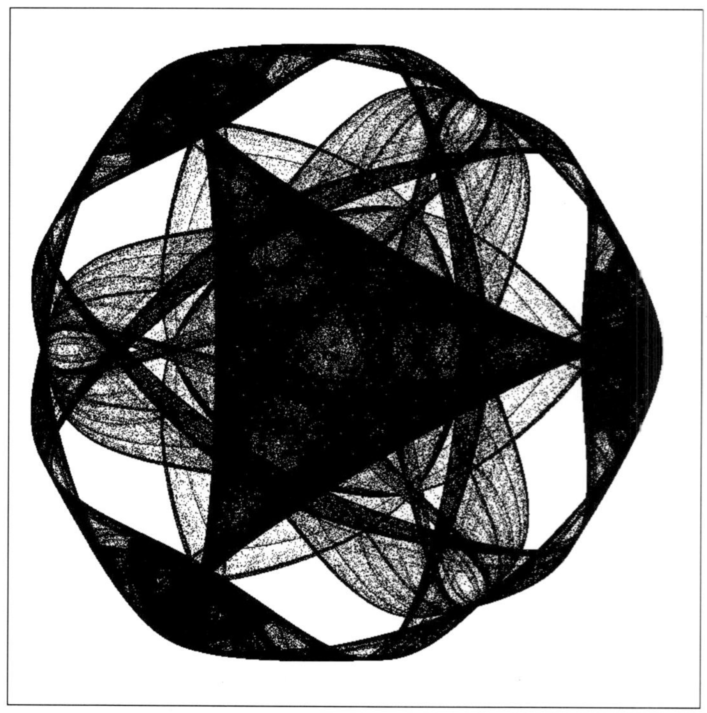

Figure 1.15 (above and bottom left) *A sequence of figures showing the effect of increasing the number of iterations.*

Figure 1.16 (bottom right) *Black and White picture with twenty million iterations.*

Figure 1.17 (opposite) *The Trampoline. A Color picture using sixty million iterations.*

ors were not discovered until relatively recently was that structure does not appear unless some type of averaging is performed. If we had recorded just 50 or 100 iterations and looked at the resulting picture, we would have been tempted to dismiss the result as just random. It was with the advent and availability of modern high-speed computers that it became possible to do many iterations and plot the resulting points in a satisfactory way. Indeed, it is worth remarking that to do just 100 iterations of the rule used in these pictures would have required more than 5000 arithmetic computations of addition, subtraction and multiplication. If we had worked by hand to seven decimal places and allowed 60 seconds for a multiplication and 10 seconds for an addition or subtraction, the time involved (assuming no errors) would have been more than 3 days! And that does not include the time that it would have taken to plot these 100 points accurately.

The pictures and discussion show that although the attractors in our pictures are symmetric, this symmetry can only be observed after a relatively large number of iterations have been made. In short, what we see is that the symmetry of these attractors is only *symmetry on average*.

What, how and why?

Until now we have spent our time trying to explain *what* the pictures we have shown are, and in general terms *how* they are made. In later chapters of this book we will address these questions in more detail. In particular, we will give the formulas that we use to make the pictures and, more importantly, show where the formulas come from. In Appendix B, we will also present computer programs that produce primitive versions of these pictures on a home computer. Now seems a good time to talk about their significance and why they might be interesting.

(a)

(b)

Figure 1.18 *(a) Fish and Eye: produced by a rule with triangular symmetry;*
(b) symmetry creation in Fish.

To be honest, the question of whether the pictures we produce and, more importantly, the mathematics behind these pictures will be of lasting scientific interest is one that has yet to be answered. We wish here to paint—with a very broad brush—some indication of why these ideas may be both interesting and important.

We begin by observing that attractors created by rules with symmetry need not have that symmetry. Figure 1.18(a) is an attractor with only bilateral symmetry produced by a rule having triangular symmetry. If we change the rule just a little bit, we find the attractor in Figure 1.18(b) which does have triangular symmetry, at least on average.

The fact that the long-term behavior of symmetric equations need not be symmetric has long been known to be of fundamental importance in physics. The book *Fearful Symmetry: Is God a Geometer?* by Stewart and Golubitsky describes many fields where this observation has proved its importance. Each of us carries an example of *symmetry-breaking* with us whenever we go for a walk. Think of the human animal walking down the street; to a very high degree of approximation we are bilaterally symmetric. Yet, what would we look like if our motion preserved that bilateral symmetry? What indeed would we look like if we were all seen to be hopping down the street moving both legs in unison, rather than walking *normally*? Normal walking breaks bilateral symmetry.

As another example, think of boiling water. We know that water never boils while you're looking at it—so don't look at it, just think about it. A typical pot in which you boil water is circularly symmetric, yet when the water begins to move as it boils, it does not move in circular waves.

A third example: starfish grow from spherically symmetric cells—why? There are many such questions and it is known that fundamental changes occur to solutions of equations when those solutions change their symmetry. Usually such changes are associated with the *loss* of symmetry, the increase in pattern, and the spawning of complicated dynamics.

The question that these pictures suggest is: can a corresponding importance be attached to the symmetry of strange attractors? In some general sense, strange attractors do represent the long-term behavior of equations, and so it is believable that the symmetry of these attractors is important. As one might imagine, however, the answer to this question is quite complex. After all, even the symmetry of strange attractors is not exact: it exists only on the average. So in order even to begin addressing the question of whether the symmetry of strange attractors is important, we have to look for examples in nature where the symmetry of a state is apparent only on *average*.

Patterns and turbulence

One area where averaged properties are known to be important is in fluid mechanics—in *turbulence*. To illustrate the issue, let us start with a thought experiment. Imagine that you are sitting on the bank of a river in flood. If you look at the surface of the river, you will see an apparently unrelated sequence of whirls, vortices, eddies, and other phenomena. At first sight there seems to be little more to say except that the events seem to occur more or less randomly. Now imagine that you take your camera and over a long period of time you take a picture of the same area of the river, say one picture every second. Take the resulting sack of slides and form the average picture. You can imagine stacking all the slides one on top of another and shining a strong light through the resulting column of slides. What you see is the 'averaged' picture of the turbulence.

It is a reasonable guess that the picture will *not* be uniform. Indeed, by carrying out this picture averaging process, you might expect to see some underlying structure to the turbulence. For example, if there was a large rock under the surface of the river or a bend in the river, this might show up as some lack of uniformity in the averaged picture. In other words, even though the motion of the river may appear random and turbulent, when you average over time, structure may

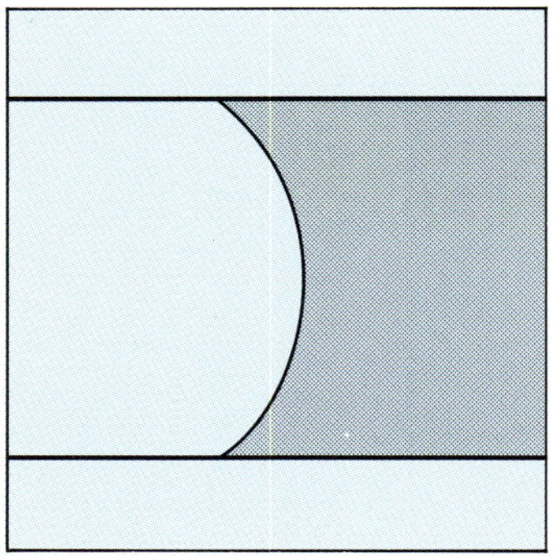

Figure 1.19 Velocity profile for laminar flow.

appear. Of course, this type of structure is not usually visible to us directly since when we see things we see them at each instant of time, and our brain is usually unable to produce a 'time-averaged' picture.

Guided by our thought experiment, we now take a look at a problem that we can discuss in a more scientific way: the flow of water down a cylindrical pipe. The water is driven down the pipe by pressure and we start by supposing that the pressure, and hence the flow rate, is small. When the flow rate is small what we see is called *laminar* flow: the molecules of water move down the pipe along lines parallel to the walls. The molecules near the center move fastest, while those on the pipe wall are stationary. In Figure 1.19, we graph the dependence of the speed of the water molecules on their distance from the pipe wall.

When the pressure (and hence the flow rate) is large, the pattern of the flow down the pipe becomes much more difficult to describe and is called *turbulent*. At relatively low flow rates, patches of turbulence pass down the pipe; at higher pressures apparently featureless turbulence is seen. Suppose one carries out a similar averaging process to the one carried out in the previous thought experiment. In particular, one can measure the speed of the fluid as it crosses a section of the pipe and compute a time average of this speed. What should we expect to see in the time average over a long period of time? The answer is contained in Figure 1.20—as we now explain.

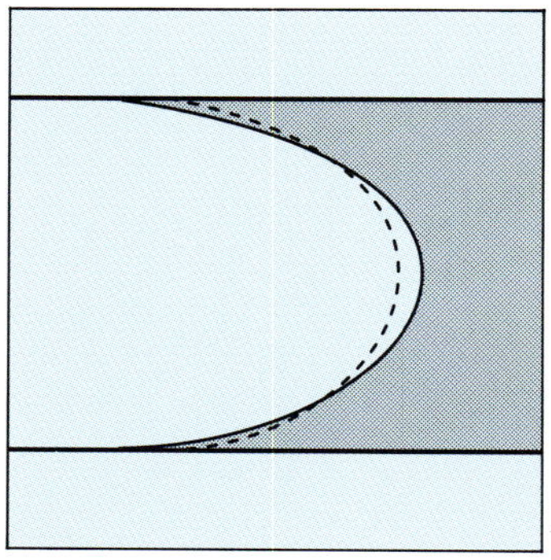

Figure 1.20 Averaged velocity profile for turbulent flow.

If, at the higher flow rate, the flow down the pipe had been laminar, then we would have expected the profile in Figure 1.19 to become elongated—indicating that the molecules in the center of the pipe were moving faster. We graph this hypothetical profile in Figure 1.20 by a solid line. The profile for the actual turbulent flow is shown with a dashed line. We see that the averaged velocity profile for the turbulent flow is flattened near the center when compared with the velocity profile for the laminar flow. What this indicates is that the higher velocity fluid at the center is intruding into regions near the wall signalling the presence of turbulence and the breakdown of laminar flow. Without the turbulence more water would flow through the pipe in a given time. Although we haven't solved the problem of designing a pipe which will stop turbulence occurring, we can see that forming the averaged picture does give us some information about properties of turbulence.

Finally, imagine what might happen if we performed this averaging experiment for turbulent flow down a pipe with square cross-section. It is a reasonable guess that we would see an averaged picture with much structure and that the picture would have square symmetry. It is conceivable, however, that at low flow rates, when the flow has just gone turbulent, we might find a picture that was *not* square symmetric. Indeed, there might be something rather curious going on when this hypothetical transition from asymmetry to symmetry takes place. At this time, it is still not known whether asymmetric turbulent flow can occur in a square pipe—but it is an intriguing question.

The Couette–Taylor experiment

You might well say 'That's interesting, but is there really any evidence that turbulent fluid flow can behave asymmetrically?' The answer to this question is an emphatic yes. There is a famous fluid dynamics experiment known as the Couette–Taylor experiment. In this experiment a fluid (usually water) is contained between two concentric cylinders with (in the simplest of the experiments) the inner cylinder free to rotate. There is an interesting change in the pattern of the fluid flow as the speed of the inner cylinder is increased. When that speed is small, the flow is laminar and the motion is called *Couette* flow. Couette flow looks the same when you rotate the cylinder about its axis and even when (in principle) you translate the flow along its cylindrical axis. When the speed

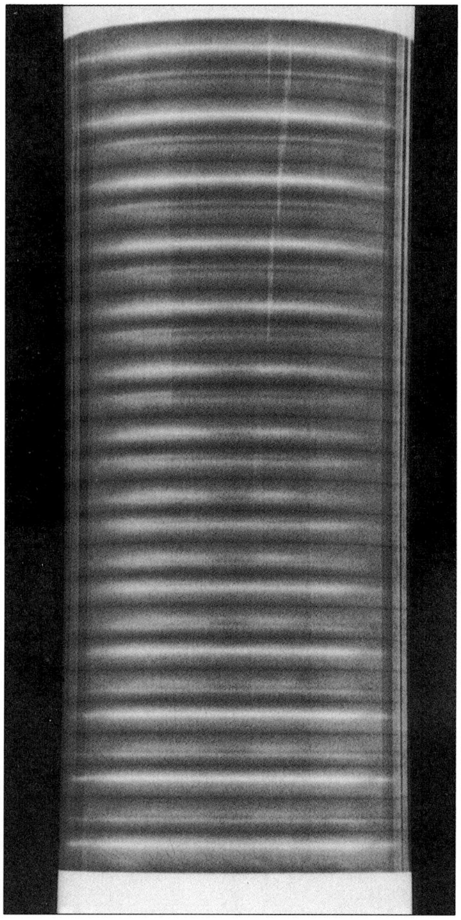

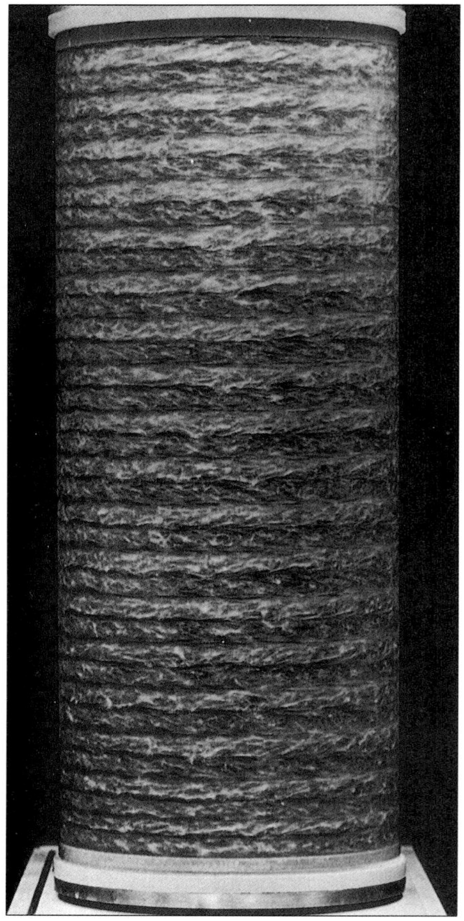

Figure 1.21 (left) *Taylor vortex flow. Courtesy of Randy Tagg and Harry L. Swinney.*

Figure 1.22 (right) *Turbulent Taylor vortices. Courtesy of Anke Brandstater and Harry L. Swinney.*

of the inner cylinder is increased Couette flow loses stability and is replaced by a flow with a most interesting pattern called *Taylor vortex flow* (see Figure 1.21). Indeed, it was a *tour de force* when in 1923 G. I. Taylor showed both experimentally and theoretically that this flow existed, and that his theory accurately predicted the value of the speed where the transition from Couette flow to Taylor vortex flow occurs. This transition is now known to be another example of symmetry-breaking: the Taylor vortex flow is less symmetric than the laminar Couette flow. Observe that Taylor vortex flow is still invariant under rotation about the axis of the cylinder, but it is no longer the same under all translations along the cylinder axis.

Experiments during the past sixty years have established the complexity of this relatively simple fluid mechanics experiment. Indeed, this experiment was the scene of one of the major verifications of the existence of chaotic dynamics. As the speed of the inner cylinder is increased, the flow becomes more and more complicated. First of all, one sees the appearance of waves, corresponding to time periodicity, then more complex motion develops, and finally, when the cylinder is spun fast enough, the flow becomes turbulent. Experimentalists Jerry Gollub and Harry Swinney were the first to observe carefully this transition to turbulence at about the same time that theoreticians David Ruelle and Floris Takens were developing a mathematical route to chaos involving just these kinds of transition.

But, from our point of view, interest in the Taylor–Couette apparatus does not end there. For if one continues to increase the speed of the inner cylinder, one arrives at a fluid state called *turbulent Taylor vortices*. As can be seen in Figure 1.22, this flow has the same general pattern as Taylor vortex flow. The motion in this flow, unlike that of Taylor vortex flow, is turbulent or chaotic. Moreover, if you look closely, the pattern in turbulent Taylor vortex flow is only approximate, though it seems to be exact *on average*. Here we have a fluid mechanical example where both patterned and unpatterned turbulence exists. And, moreover, the symmetry that appears in the turbulent states occurs only on average.

? Couette 1 → Taylor 2 → Turbulent 3 → Turbulent Taylor 4
Change in phase from 1-2 is similar to that from 3-4

Some further speculation

In our discussion of turbulence, we have created thought experiments about the flow of fluid in a river or in a square pipe where patterns and symmetry on average might be detected. We have described an actual fluid dynamics experiment where a pattern in a turbulent flow is known to exist, at least on average. It would seem fair, then, to ask: 'Where would we see the effects of symmetry on average in the *real* world'? To address this point, we present another thought experiment.

Imagine a two-dimensional rectangular container that holds a fluid mixture of several chemicals, one of which is an acid. Imagine also that the fluid mixture is in motion, indeed in complicated chaotic motion. Finally, imagine that the acid can eat away or etch the bottom of this container. Under these circumstances it is reasonable to presume that the rate of etching along the bottom of the container will be proportional to the concentration of the acid. Of course, that concentration may be expected to vary both along the bottom of the container and in time, and the etching pattern created over time can be expected to be quite complicated.

There is one nontrivial symmetry in the thought experiment we have posed. The container can be reflected left to right without making any changes in its description. With this symmetry in mind, we ask the question:supposing the fluid circulates for a long time, will the pattern of etching along the bottom of the container be left–right symmetric or not? We expect that both possible answers—yes and no—will occur depending on the particular system that is studied. Moreover, we expect there to be critical parameter values of the system where the etching pattern will suddenly jump from being asymmetric on average to being symmetric on average when these parameters are changed.

It is important to understand that we are not saying that the dynamics of the concentration profile of the acid along the bottom will be symmetric at each point in time—it will not be. What we are saying is that if one averages this concentration in time over a relatively long period, then what one will see in some systems is an averaged concentration with a distinctive left–right symmetric pattern, even though the concentration is changing chaotically in time. In this sense, the pattern will appear only on average.

To actually make a detailed study of this kind of fluid motion, with its associated chemical reactions, is a very complicated task.

Rather than embark on such a complex project straight away, it makes sense first to study simpler problems to see whether the phenomenon of symmetry and pattern on average is present. The simplest possible model of such an experiment would be one that keeps track only of the concentration of the acid along the bottom of the container. Equations of this sort are called partial differential equations and describe the evolution of a quantity like concentration in both space and time. The problem with this kind of simplification is that it is hard to write down an equation that can in any sense be called approximate.

Another approach is just to choose a partial differential equation that is used to model a chemical process, solve it numerically, compute the time average, and see whether or not the time average is symmetric. This has been done recently for a model equation called the Brusselator. (It was written down by a group of chemists in Brussels; there is also a more complex model called the Oregonator written down —you've guessed it—by a group in Oregon.) This numerical experiment was performed by Michael Dellnitz; what was found was a transition from an asymmetric time average to a symmetric one as a parameter was changed. In our thought experiment, this parameter would correspond to the length of the bottom of the container.

This kind of numerical experiment allows us to believe that the phenomenon of symmetric patterns on average is real. We can now begin to wonder whether other phenomena where distinctive (almost) symmetric patterns appear, such as in the shells of sea creatures, are the product of chaotic growth, averaged to form a regular pattern.

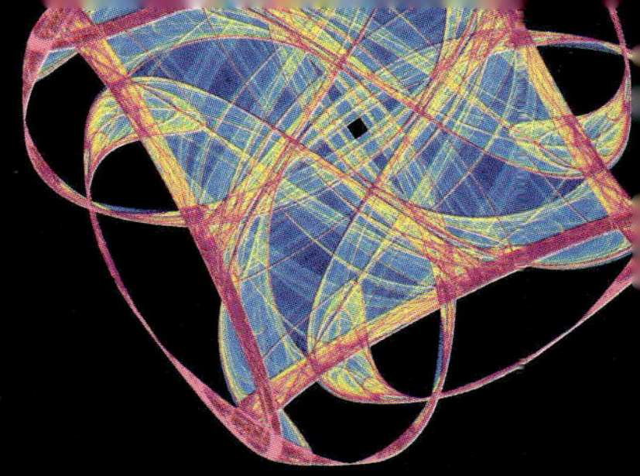

Chapter two

PLANAR SYMMETRIES

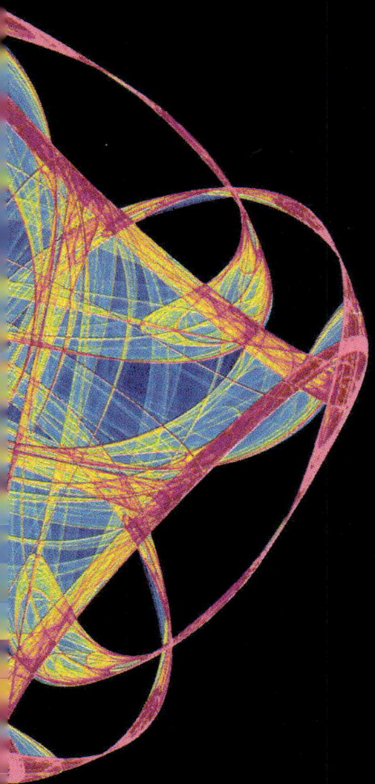

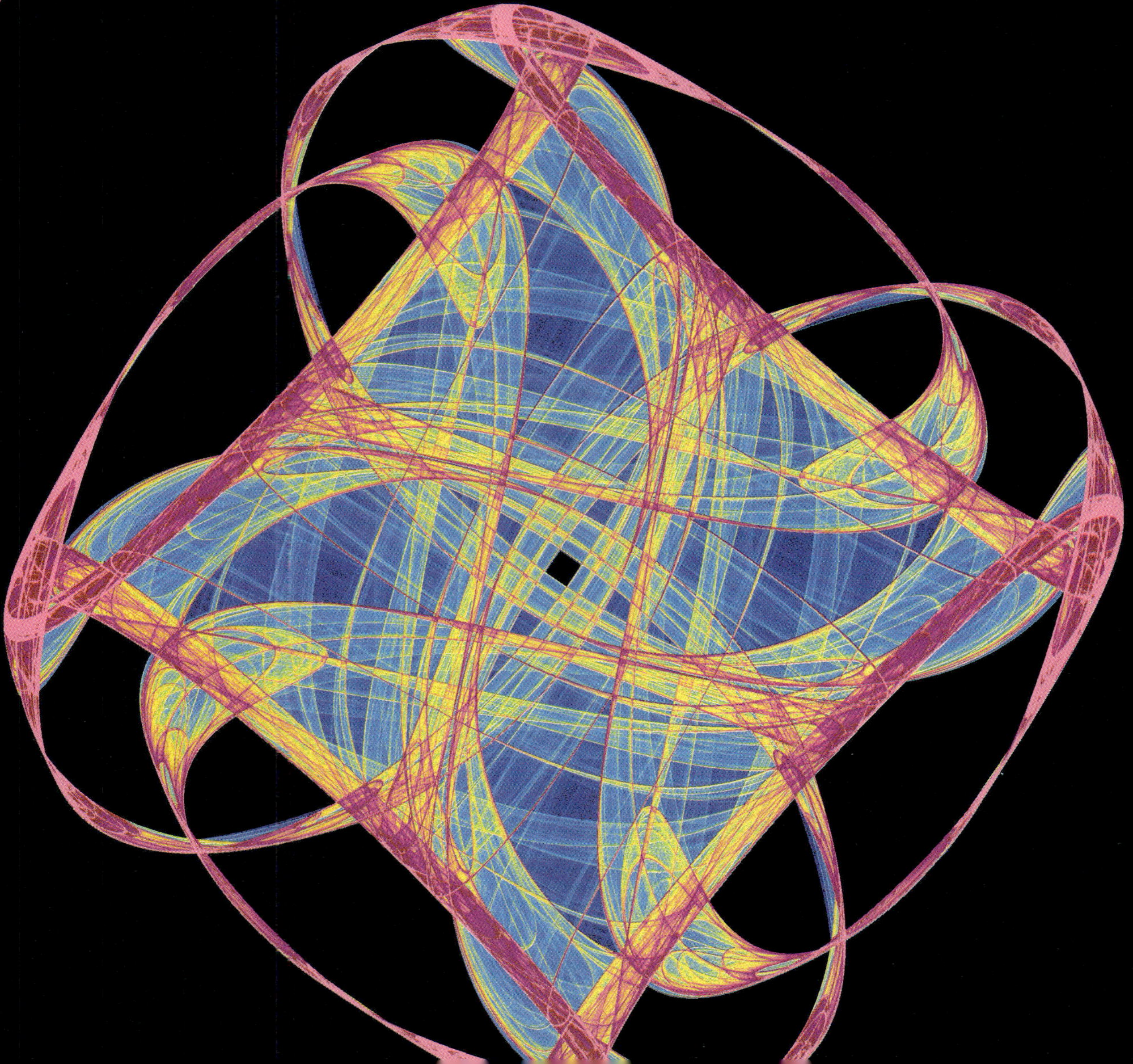

SYMMETRY seems such a basic notion that it should be possible to go to a standard reference and find out what it is. With this in mind we consulted the well-known *Mathematics Dictionary* edited by James and James and found the following rather perplexing entry under *symmetry*: 'See various headings under SYMMETRIC'. And under *symmetric*: 'Possessing symmetry'. In fact, there are two difficulties that have conspired to complicate our seemingly reasonable request, and have led to these circular references.

The first difficulty is one of abstraction. In our introductory chapter we spoke of symmetries as individual motions that leave a picture unchanged. In mathematics what is important, however, is not the individual symmetries, but rather the collection of all symmetries of the picture. This collection is called a *group*, and much of the discussion in this dictionary refers to group theory. The second difficulty concerns scope. There are so many different kinds of symmetry that it is virtually impossible to give one simple all-inclusive definition of this term. For example, one can discuss the symmetries of an equation or a physical law, not just the symmetries of a picture. Although the ideas of symmetry in its different manifestations are related, care must be taken when making the notion precise.

In our discussion of symmetry, however, our goals are really quite modest. We want to know what are the possible symmetry groups for pictures in the plane. From the perspective of a general discussion of symmetry groups, we have asked a very specialized question. Group theorists know the answer to this question, and it is this answer that we shall attempt to describe.

Types of symmetric pictures

We begin with some ground rules. In Chapter 1 we defined a symmetry of a (planar) picture to be a motion of the plane that leaves that picture unchanged. We noted, for example, that there are six symmetries of a three-pointed star: rotation counterclockwise through 120° and 240°, reflection across the three axes of the star, and the 'trivial' motion that leaves the plane fixed. For figures like the equilateral triangle and square, symmetries are always either rotations or reflections. On the other hand, when we consider repeating patterns, such as are found in wallpaper patterns and quilts, we will also have *translational* symmetries, which slide everything along without rotating.

Our pictures are all determined on the computer screen, which represents just a small part of the infinite Euclidean plane. Yet there are still two types of picture: those that appear to be contained in the computer screen and those that do not. We call the first kind of picture a *symmetric icon*. These pictures have only rotations and reflections as symmetries. Such symmetries are related to those of a regular n-sided polygon, such as an equilateral triangle, a square, or a regular pentagon. The second kind of picture appears in our minds to fill the whole Euclidean plane; these pictures are determined by infinite repetition from the small part of the plane that we see on the computer screen. We call these pictures the *quilt patterns*; all of these pictures have translational symmetries. Thus we see that the two types are distinguished by whether or not they have a proper translation as a symmetry. In our discussion, we first describe the symmetry groups of the icons, that is, those pictures that have only rotational and reflectional symmetries.

Groups

There are three fundamental properties of the collection of symmetries of any picture, and it is these properties that mathematicians use in their analysis of symmetry groups.

The first seems so obvious that it hardly seems worth stating: there is a trivial symmetry that we denote by I. This trivial symmetry keeps the picture fixed and plays the same role for symmetries that the number zero plays for integers.

The second rule is almost equally clearly valid. Any symmetry has an equal but opposite symmetry, called the *inverse symmetry*. If rotation counterclockwise through the angle $d°$ is a symmetry of a picture, then rotation clockwise through $d°$ is also a symmetry. If translation in one direction is a symmetry, then so is translation in the opposite direction.

Finally, performing two motions one after the other is also a motion, called the *composition* of the first two motions. The third property states that if the first two motions are symmetries of a picture, then so is the composition.

These three properties actually define an abstract *group*; this is the reason that we call the collection of all symmetries of a picture a symmetry group. We discuss these ideas in more detail below in terms of the symmetry group of the square.

Symmetries of regular figures

In his autobiographical book *Surely You're Joking, Mr. Feynman!*, the physicist Richard Feynman describes the behavior of mathematicians at Princeton in the following terms. A mathematician would try to describe a new idea to one of his colleagues. After much discussion, puzzled looks and many filled blackboards, his colleague would eventually say, 'Oh I understand, *that's* trivial!' To non-mathematicians, who have wrestled with high-school mathematics, this comment may seem quite bizarre. Yet there is some truth in the assertion that much of mathematics is inherently trivial or obvious. In part, this is because questions in mathematics generally admit a yes–no answer. Put another way, if you make a mathematical assertion, then it is either true or false. This simple situation is to be contrasted with the everyday problems of belief and action that we all confront. Such everyday, and apparently mundane, questions are in reality very complex precisely because they cannot be reduced logically to a matter of simple truth or falsity. Even in science, there are no absolute truths, only reasonable approximations to what is tacitly believed to be an underlying, yet invisible, truth. However, just because a piece of mathematics is 'obvious' does not mean that it is easy to understand. Indeed, the most obvious ideas are often the most difficult.

As an aid to understanding our discussion of symmetry, we suggest that you get pencil, paper, and scissors and test out what we say while reading through this section. By the end, we hope that everything we do will indeed be obvious!

We begin our discussion of the symmetry group of a regular n-sided polygon with that of the square.

Symmetries of a square

It's fairly easy to convince yourself that there are precisely eight symmetries of a square. First there is the trivial *identity* symmetry got by picking up the square and putting it back down exactly as it was before. We denote this symmetry by I. Then there are the three rotations of the square obtained by rotating the square counterclockwise by 90°, 180°, and 270°. These symmetries are denoted by r_{90}, r_{180}, and r_{270}.

Note that r_{180} and r_{270} can be obtained by applying r_{90} twice and three times respectively. Indeed, applying r_{90} four times just gets back to the identity I. Finally, we can flip the square over on itself while leaving the vertical midline of the square fixed. We denote this motion by F. The remaining three symmetries of the square are now obtained by flipping the square using F and then rotating the square using one of three rotations listed above.

We denote the group of all eight symmetries of the square by $\mathbf{D}_4$—the $\mathbf{D}$ standing for *dihedral* and the 4 standing for the regular four-sided polygon (the square). We also note that the identity together with the three rotations form a group called the *cyclic group* $\mathbf{Z}_4$.

It is worthwhile being a little more explicit about the motions that we have just described. One way to do this is to cut a square out of cardboard and color the opposite sides of the square with different colors, say black on one side and white on the other. Place the square on a table, with the white side uppermost, and label the corners A, B, C, and D counterclockwise around the square. Turn the square over, and label the corresponding corners A, B, C, and D so that labels match up for both sides (A, B, C, and D will run clockwise round the black side of the square). Label the center of the square, on both sides, with the letter O. We show the labelled square, white side up, in Figure 2.1. Now place the square on a sheet of paper, with white side up, and draw the outline of the square on the paper. Mark the corners A, B, C, and D on the paper and then mark the center of the outlined square with an O.

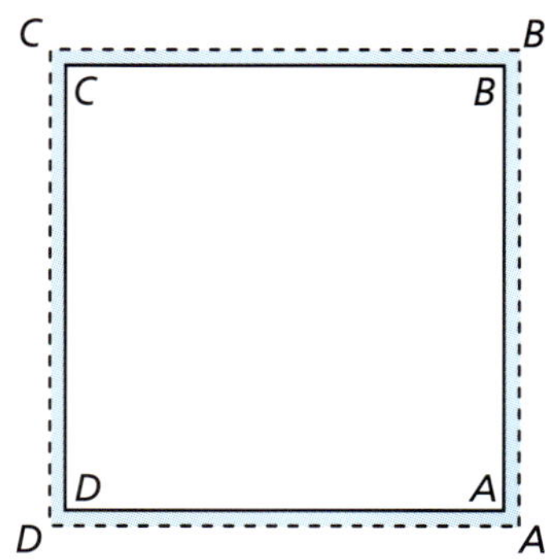

Figure 2.1 *Cardboard square, white side up.*

The first thing to notice is that there are precisely four ways to place the square within the outline with the white side up: these are the symmetries in the group $\mathbf{Z}_4$ mentioned previously. Using the A, B, C, and D designations of the corners we can describe each of the symmetries of the square. For example, the counterclockwise rotation r_{90} moves the corner labelled A to the corner labelled B, the corner B to C, C to D,

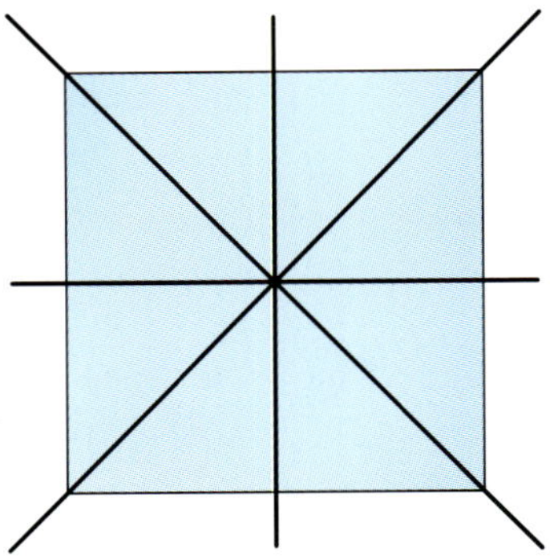

Figure 2.2 *The axes of symmetry of the square.*

and D to A. We can write this motion using the shorthand notation

$$(A, B, C, D) \rightarrow (B, C, D, A).$$

The flip F can now be described using this notation as

$$(A, B, C, D) \rightarrow (D, C, B, A),$$

that is, F interchanges corner A with corner D and corner B with corner C. It is easy to see that when you try to interchange these corners with a motion of your cardboard square you must turn the square over so that its black side is up.

As an exercise, you might want to check that the symmetry obtained by first flipping the square using F and then rotating the square using r_{90}, which we denote by $r_{90}F$, leads to the motion

$$(A, B, C, D) \rightarrow (A, D, C, B),$$

which just interchanges the corners B and D.

We can now delve a little further into the geometry associated with these symmetries. The four symmetries that involve F (namely, F, $r_{90}F$, $r_{180}F$, $r_{270}F$) all have axes of symmetry (F leaves the horizontal midline fixed while $r_{90}F$ leaves the diagonal from corner A to corner C fixed). Indeed there are four axes of symmetry of the square (see Figure 2.2) and these correspond to the four symmetries just mentioned. On the other hand, the rotations themselves have no axes of symmetry.

We can now use this discussion of the symmetries of the square to point to properties of patterns having this symmetry. For example, the fact that the eight symmetries of the square can be obtained from just two of the symmetries (namely, r_{90} and F) allows us to check whether a figure has square symmetry by just asking whether the figure remains the same when you rotate it by 90° and when you flip it by F. Of course this figure might have more than square symmetry (a regular octagon has square symmetry, and more) but it has at least $\mathbf{D}_4$ symmetry.

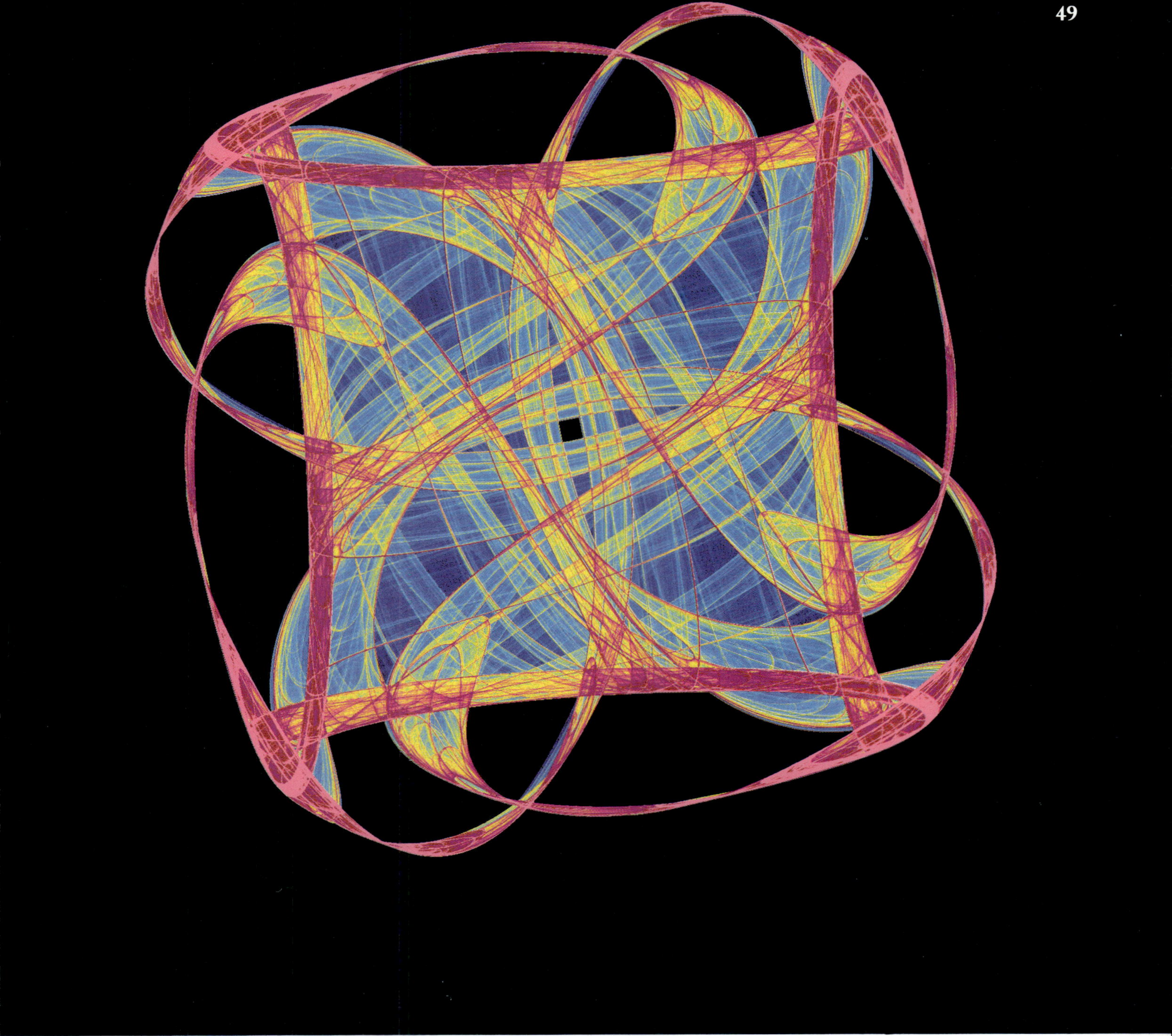

Figure 2.3 Swirling Streamers. An Icon with rotational $\mathbf{Z}_4$ symmetry.

Suppose a figure has the rotational symmetry r_{90} but does not have the flip symmetry F. From the previous discussion we say that this figure has $\mathbf{Z}_4$ symmetry but not $\mathbf{D}_4$ symmetry. We can also make a prediction about that figure: it will *not* have any axes of symmetry. Indeed, look at the picture in Figure 2.3 which has this property.

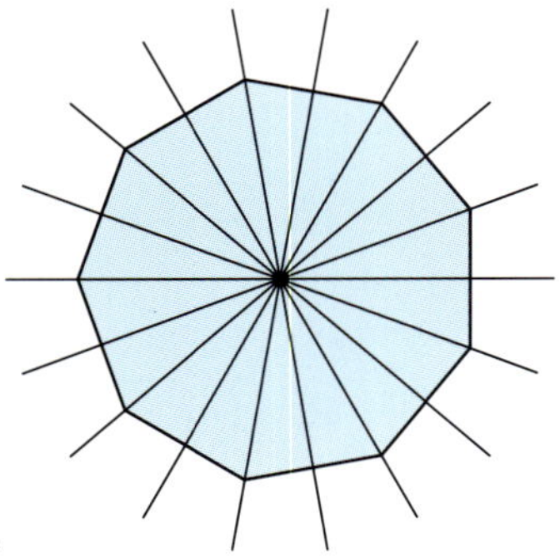

Figure 2.4 *The axes of symmetry of the regular nonagon.*

Symmetries of a regular polygon

Everything we have said for the square extends to the symmetries of the other regular figures in the plane. For example, suppose we consider the regular nine-sided polygon: the *nonagon* (see Figure 2.4). There are nine rotational symmetries of the regular nonagon forming a group denoted by $\mathbf{Z}_9$ (the cyclic group of order 9). In all, there are a total of 18 symmetries of the regular nonagon forming a group denoted by $\mathbf{D}_9$ (the dihedral group of order 18). The group $\mathbf{D}_9$ contains nine rotational symmetries and nine reflectional symmetries about the nine axes of symmetry of the regular nonagon. We may replace '9' in this discussion by any whole number 'n' greater than or equal to two. The resulting regular n-sided polygon has symmetry group $\mathbf{D}_n$ (the dihedral group of order $2n$) containing $2n$ symmetries, n of which are rotational symmetries and the remainder of which are reflections in the n axes of symmetry of the regular n-sided polygon.

We should emphasize that we have chosen regular figures for our discussion of symmetry groups simply because they are the simplest figures possessing the given symmetry and so are easier to draw and to talk about. Of course, everything we have said applies equally well to the more complex figures illustrating symmetric chaos that we show throughout the book. For example, in Figure 2.5 we show a picture with $\mathbf{D}_9$ symmetry.

We end this subsection by noting that the groups of symmetries of symmetric icons are either the dihedral groups $\mathbf{D}_n$ or the cyclic groups $\mathbf{Z}_n$. Needless to say, the corresponding discussion of the symmetries of symmetric icons in three dimensions is substantially more complicated.

Figure 2.5 (opposite) *Lace by Nine.*

Tilings and wallpaper patterns

Figure 2.6 (opposite) *Emerald Mosaic. Chaotic tiling with square symmetry.*

Although figures with the symmetry of regular *n*-sided polygons are common in many motifs, our most direct experience of symmetry is through observation of wallpaper and tiling patterns. In this section, we enlarge our discussion of symmetry to include this type of repeating pattern.

We begin with a simple tiling pattern based on the square (see Figure 2.6). This pattern is formed by simply repeating the design in one basic square, which we call a *cell*. From the perspective of symmetry, the most important feature of this pattern is the translational symmetry; the pattern remains unchanged if we translate the figure either horizontally or vertically by a cell. The existence of these translational symmetries allows us to imagine extending the quilt-like design to the whole plane, after having seen only several repetitions of the basic cell on which the pattern is based. The second, more refined, feature of this figure is the square $\mathbf{D}_4$ symmetry of the design in the basic cells.

Figure 2.7 *Tiling of the plane by squares.*

Indeed, two distinct properties of the square are used when describing this 'simple' tiling pattern. The first property is that horizontal and vertical translations of the square neatly fill up the entire plane, as indicated in Figure 2.7. In this way it is possible for translations to be symmetries of the entire pattern.

The second property based on the square is the fact that a square has square symmetry, and we can speak of the square $\mathbf{D}_4$ symmetry of the design within the unit square.

Figure 2.8 *The hexagonal tiling of a honeycomb.*

From this perspective it is easy to imagine a tiling pattern built on hexagons which also fill out or *tile* the plane, as any self respecting honey-bee knows (see Figure 2.8). Moreover, we can arrange things so that within each cell the design has hexagonal ($\mathbf{D}_6$) symmetry (see Figure

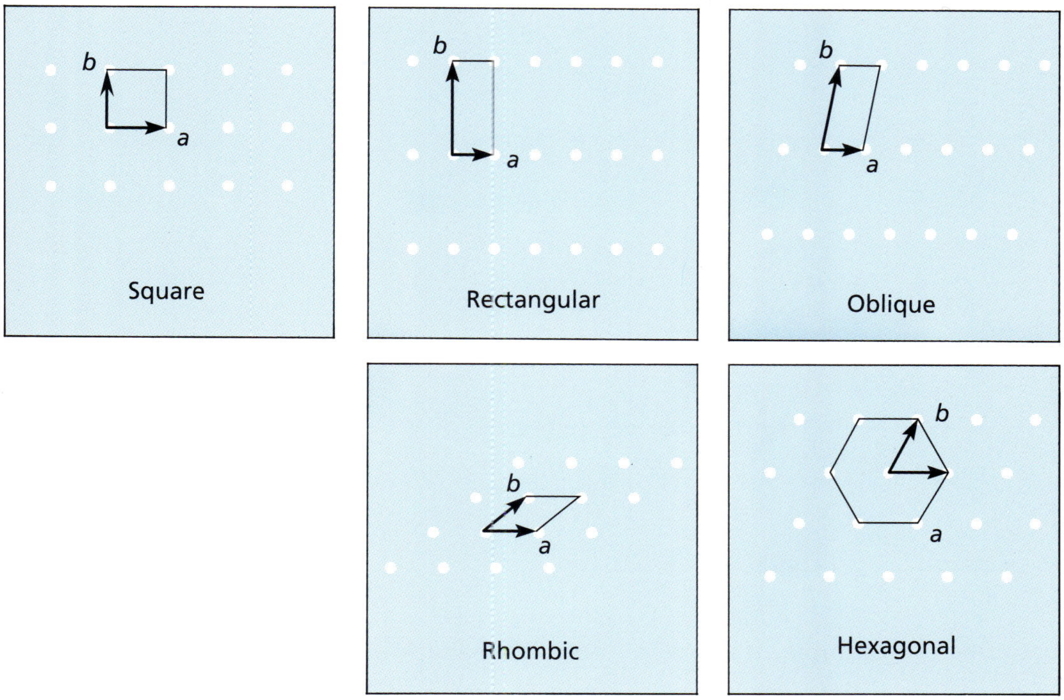

Figure 2.9 (opposite) *Dutch Quilt.*

Figure 2.10 (left) *Regular tilings of the plane.*

2.9). We refer to this tiling of the plane as a *hexagonal tiling*. It turns out that there are just five different ways to fill the entire plane by simply translating one regular tile. In addition to the square and hexagonal tilings, there are the rectangular, rhombic, and oblique tilings (see Figure 2.10). In practice, the repeating patterns that are most often seen in nature or art are based on either square or hexagonal tiles.

The wallpaper groups

Although in this volume we shall emphasize patterns built with square symmetry on square tiles and hexagonal symmetry on hexagonal tiles, it is interesting to consider other types of repeating

Figure 2.11 *Rotated A's on a square tiling.*

patterns. For example, it is easy to imagine a pattern based on a design on a square tile that is not square symmetric, but nevertheless is repeated infinitely often. We can also repeat that design by rotating it counterclockwise by 90° whenever we translate it either up or to the left and clockwise by 90° when we translate it either down or to the right (see Figure 2.11). Note that a 4 × 4 grid of cells in this figure repeats infinitely often but there is added structure within the grid. For more information on the appearance and use of these types of design in different cultures, we suggest that you consult the book *Symmetries of Culture* by Dorothy Washburn and Donald Crowe.

The collection of all such symmetry patterns, based on tilings in the plane, have been classified by their symmetries; these symmetries are called either the *wallpaper groups* or the two-dimensional *crystallographic groups*. There are seventeen different wallpaper groups, and hence seventeen different types of wallpaper pattern. It is an interesting exercise to try to find all of the seventeen different patterns. In fact, it is

Figure 2.12 Sugar and spice.

quite tricky to show that all of these patterns are actually different and that there are no other patterns.

We end this section with two pictures of tiling patterns (Figures 2.12 and 2.13) that are computed using mathematical rules that have the symmetry of square and hexagonal tilings. These pictures indicate the rich variety of patterns that can be obtained with chaotic tilings.

Figure 2.13 (opposite) *Crown of Thorns. A hexagonal quilt.*

Coloring and interlacing

In practice, of course, patterns must be drawn using at least two colors and many ornamental patterns are drawn using a complicated set of lines and curves. Both of these observations lead to more sophisticated classifications of pattern in ways that we now discuss.

We begin with color. For simplicity, suppose we use two colors, say *black* and *white*. We call the pattern *positive* if it is drawn black

Figure 2.14 *Symmetry drawing A by Maurits Escher from The Graphic Work of M.C. Escher.*

Figure 2.15 *Interlace pattern from the Dome of the Rock in Jerusalem*

on white and *negative* if it is drawn white on black. We can now imagine making a design, such as the one by Maurits Escher (Figure 2.14), by taking a square tiling pattern and alternating positive and negative pictures, as on a chessboard. (In Escher's picture the negative image is reflected.) It turns out that there are exactly forty-six of these alternating two color tiling patterns (for details, we again refer you to the book by Dorothy Washburn and Donald Crowe). We resist the temptation even to discuss the symmetries of three-color tiling patterns, though that classification has also been completed!

As we noted, ornamental designs are often constructed using complexes of lines and curves. In Figure 2.15, some lines intersect while others pass each other in a way reminiscent of a weaving pattern. The effect is similar to what one sees in a detailed street map where intersections, overpasses, and underpasses are marked. Hermann Weyl, in his book *Symmetry*, devotes a chapter to the subject of ornamental patterns.

Chapter three

PATTERNS EVERYWHERE

SYMMETRY stands out, it demands our attention. Viewed in this light, it is not suprising that symmetry plays such a pivotal role in a wide range of human activity, from decorative design and textiles to architecture and advertising logos. Yet symmetry means much more to us than just a pleasing visual response. For example, symmetry has enjoyed a special place in both philosophy and theology. In this context, symmetry is the great harmonizer and unifier. We frequently portray, obtain, or symbolize, *meaning* by the use of symmetry. Well-known instances of this can be found in Greek philosophy ('harmony of the spheres'), in the religious icons and stained glass of Christianity, and, most especially, in the exquisite mosaics and patterns used in Islamic art. This process of using symmetry to obtain harmony and unity is also to be found in science. Many of our modern ideas about the universe are based, just as those of the Greeks were, on symmetry.

Our response to symmetry is more complex than might at first appear. The most symmetric figure in the plane, bar the plane itself, is the circle. Even though the Greeks regarded the circle as perfect—being most symmetric—we tend not to regard objects with the symmetry of the circle as being especially attractive. Indeed, in this book we have not shown any symmetric icons with full circular symmetry on the grounds that they look rather dull when compared with the figures with dihedral or cyclic symmetry. We respond best when symmetry is broken from the perfect symmetry of the circle (or plane) to a lesser, but still identifiable, symmetry. When we break the symmetry of the circle to that of the square, we lose the sameness or homogeneity of the circle and find the structure of the square.

This structure is comprehensible, even when it is represented by a symmetric (chaotic) icon with square symmetry. The squareness allows us to encompass it within our vision and mental capabilities.

Once all symmetry has gone, we often lose control, and words and metaphors are no longer adequate to describe what we see. However hard we try, it is difficult to impose order and rationality on the unsymmetric world around us, whether it be nature, the stock market, the weather, or humanity itself.

In Chapter 1, we pointed out that the word *chaos* has two distinct meanings. We agreed to follow contemporary scientific usage by deciding that chaos meant 'complex and unpredictable'. However, the original meaning of chaos is that of 'being without feature or form'. We shall refer to this state as one of *total chaos*.

Perfect symmetry and total chaos have one feature in common: both look the same at every point and from every direction. In this sense, total chaos can be thought of as perfectly symmetric. Symmetry, however, when used in art, decorative design or architecture, is usually one step down from perfect symmetry. Analogously, our pictures of symmetric chaos can be viewed as a breaking of the perfect symmetry of total chaos. By breaking this symmetry, we introduce (symmetric) structure. If we think of chaos as representing the 'dark side', then one way we can illuminate the otherwise invisible symmetry and structure in chaos is by the use of time averaging.

Throughout this book, we show a range of computer-drawn pictures with various symmetries. We have found that these pictures evoke in many people memories of images from a wide range of human experience. Of course, likenesses are rarely exact, but they *are* often highly suggestive. In a sense, the pictures seem to act like a

Rorschach ink-blot test, eliciting responses from *Star Wars* to starfish. It is difficult to explain exactly why this is so.

We would like to think that one reason is that the symmetric creations of humanity, obtained by breaking perfect symmetry, are similar in kind to those we obtain by breaking the symmetry of total chaos.

In this chapter, we have collected a number of images from nature and art and compared them with similar computer-drawn pictures. In no sense are we suggesting that the mathematics used to draw these pictures also provides an explanation for why the images look the way they do. Specific resemblances, such as they are, are surely accidental. What is surprising is that the character of these pictures, constructed using symmetric chaos, is so close to the character of symmetric images from nature and art. A consequence is that it seems possible that computer-drawn pictures, based on the ideas of symmetric chaos, can provide images that will be useful in design. At any rate, we feel that there is a beauty in these pictures of symmetric chaos that is worth recording.

Icons with pentagonal symmetry

Let us start by examining the classification of images by their symmetry in a little more detail. We know that there are many different kinds of images that have the same symmetry. To illustrate this point, we focus on images with pentagonal or $\mathbf{D}_5$ symmetry. There are pictures of diatoms (Figure 3.1), flowers (Figure 3.2), corporate logos (Figure 3.3), architectural designs (Figure 3.4), and sea creatures (Figure 3.5). Along with certain of these images we have included examples of computer-drawn pictures that appear to mimic the real images. Of course we should not forget the ultimate in logos with pentagonal symmetry: the Pentagon itself (Figure 3.6).

Figure 3.1 (above left) *A unicellular diatom with pentagonal symmetry.*

Figure 3.2 (*a*) (above right) *The Spring Gentian and* (*b*) (right) *The Common Mallow.*

Figure 3.3 *Corporate logos in a Texas beach scene. Original watercolor by Wendy Aldwyn.*

You may wonder what some of these images actually are. So here is some background material. Our examples of marine life include a type of diatom *Actinoptychus heliopelta* that occurs in marine and freshwater plankton. Most flowers possess dihedral symmetry and there are many examples of flowers with five petals and fivefold symmetry. This really is a case where the symmetry is there to be noticed—in this case by the bees. We have chosen the common mallow (*Malva sylvestris*) and the spring gentian (*Gentiana verna*) to illustrate two of the many possible choices.

For corporate logos we have chosen the Chrysler automobile logo, since, as it happens, we have been driving Chrysler cars for the past few years and, having stared through that logo on many a long trip, we feel justified in using it. Our architectural design comes from the Gothic tracery at the Cloisters of Hauterive. Finally we end our discussion of pentagonal symmetry with pictures of starfish and sanddollars that are found in great abundance along the Texas coast. The computer-drawn sanddollar bears an uncanny resemblance to the real shell.

Figure 3.4 (On page 68)
(*a*) (below) *Gothic tracery from the Cloisters of Hauterive and*
(*b*) (above) *Gothic Medallion.*

Figure 3.5 (On page 69)
(*a*) (below) *Starfish, sea cookies, and sanddollars and* (*b*) (above) *The Sanddollar.*

Figure 3.6 (On page 70)
(*a*) (below) *The Pentagon and*
(*b*) (above) *The Pentagon Attractor.*

Figure 3.7 (On page 71)
(*a*) (below) *St. John's wort and*
(*b*) (above) *Chaotic Flower.*

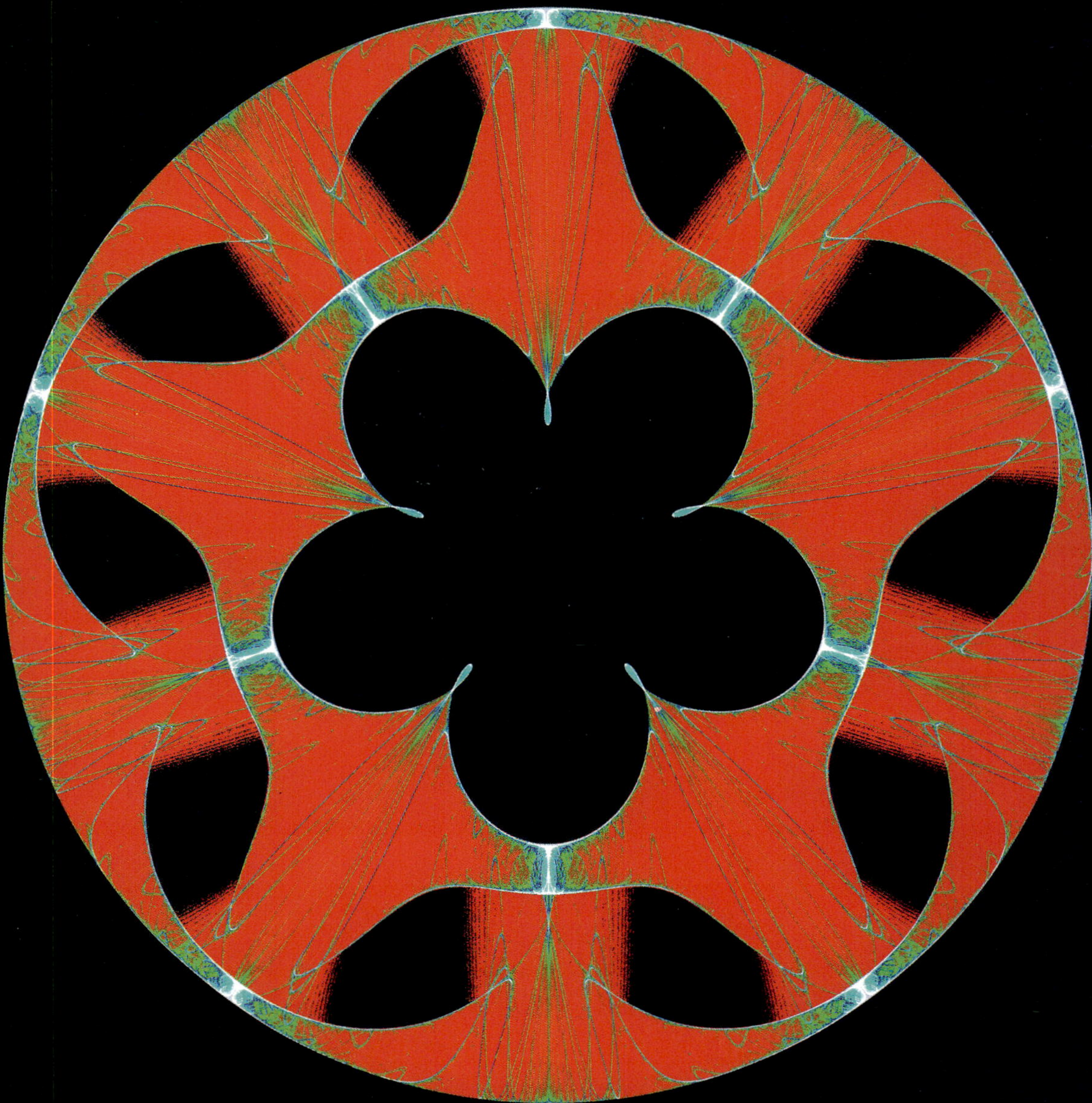

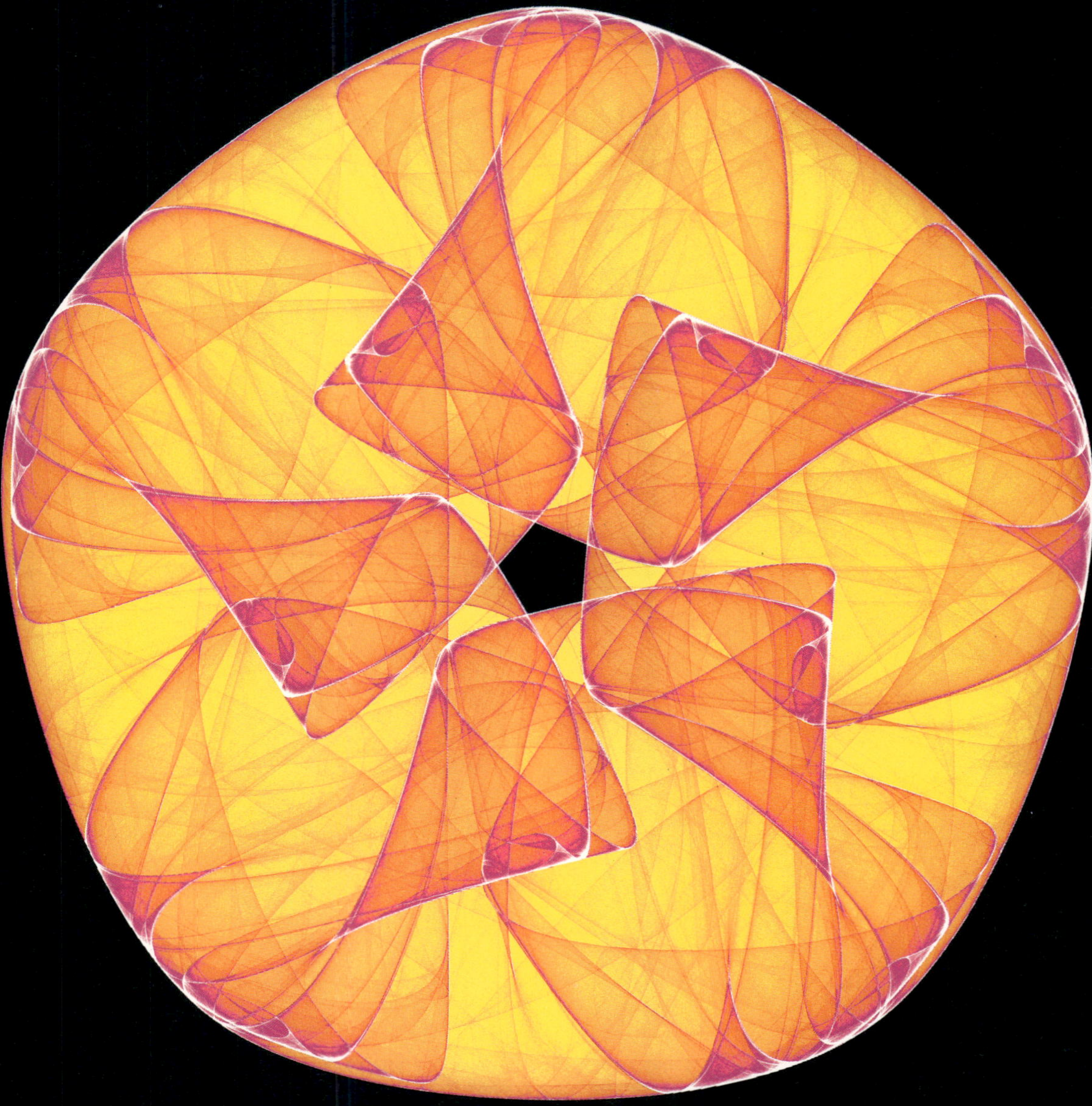

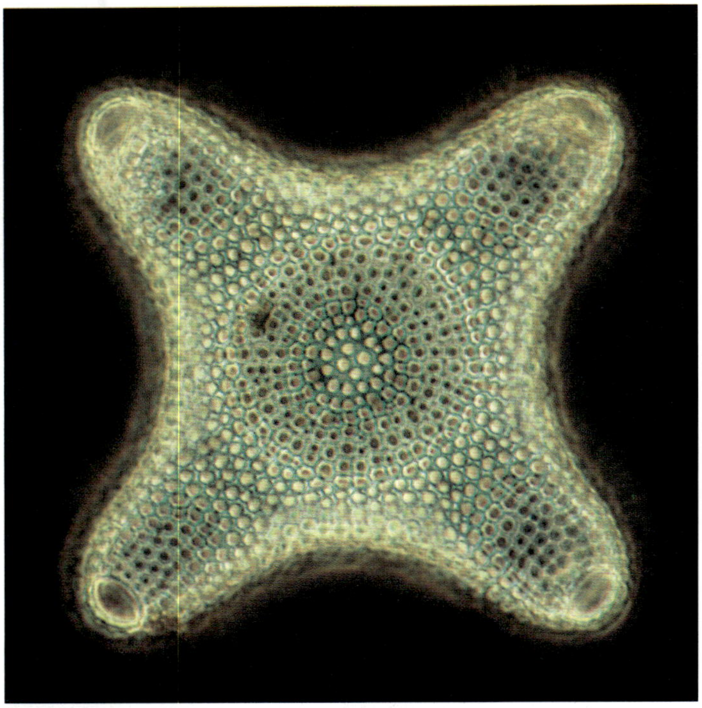

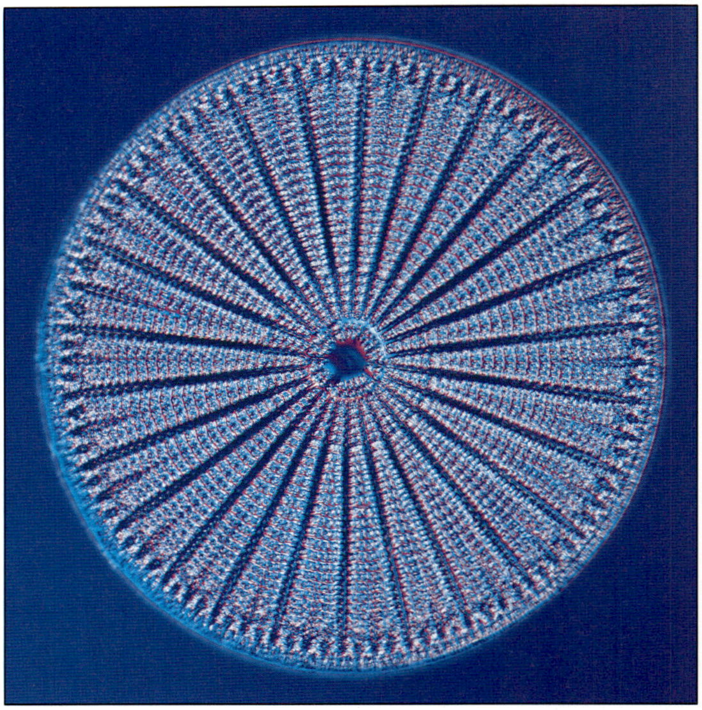

Figure 3.8 *Diatoms with* (*a*) (above left) *square and* (*b*) (above right) *26-fold symmetry.*

Icons with different symmetries

There are many images with different kinds of symmetry that are possible to collect. For example, although most flowers have dihedral symmetry, the St John's wort (*Hypericum perforatum*) (Figure 3.7) has cyclic $\mathbf{Z}_5$ symmetry. Unusual symmetries suggest possible unusual uses and, indeed, in the *The Kindly Fruits* we find that the St. John's wort is used

to cure madness, especially if the patient was thought to be possessed by the devil, and was collected on St John's Eve (23 June) to be hung in the house to ward off evil spirits.

There are diatoms with $\mathbf{D}_4$ and $\mathbf{D}_{26}$ symmetry pictured in Figure 3.8. The Mercedes-Benz symbol in Figure 3.9(a) provides a good example of a logo with triangular symmetry; the Mitsubishi corporate logo pictured in Figure 3.3 is another example.

Figure 3.9 (*a*) (oposite below) *The Mercedes-Benz medalion and* (*b*) (opposite above) *The Mercedes-Benz Attractor.*

Distinctive logos can generate powerful images and associations, as is evidenced by the following quote about the Mercedes medallion from Roger Bell's introduction to *Great Marques: Mercedes-Benz.*

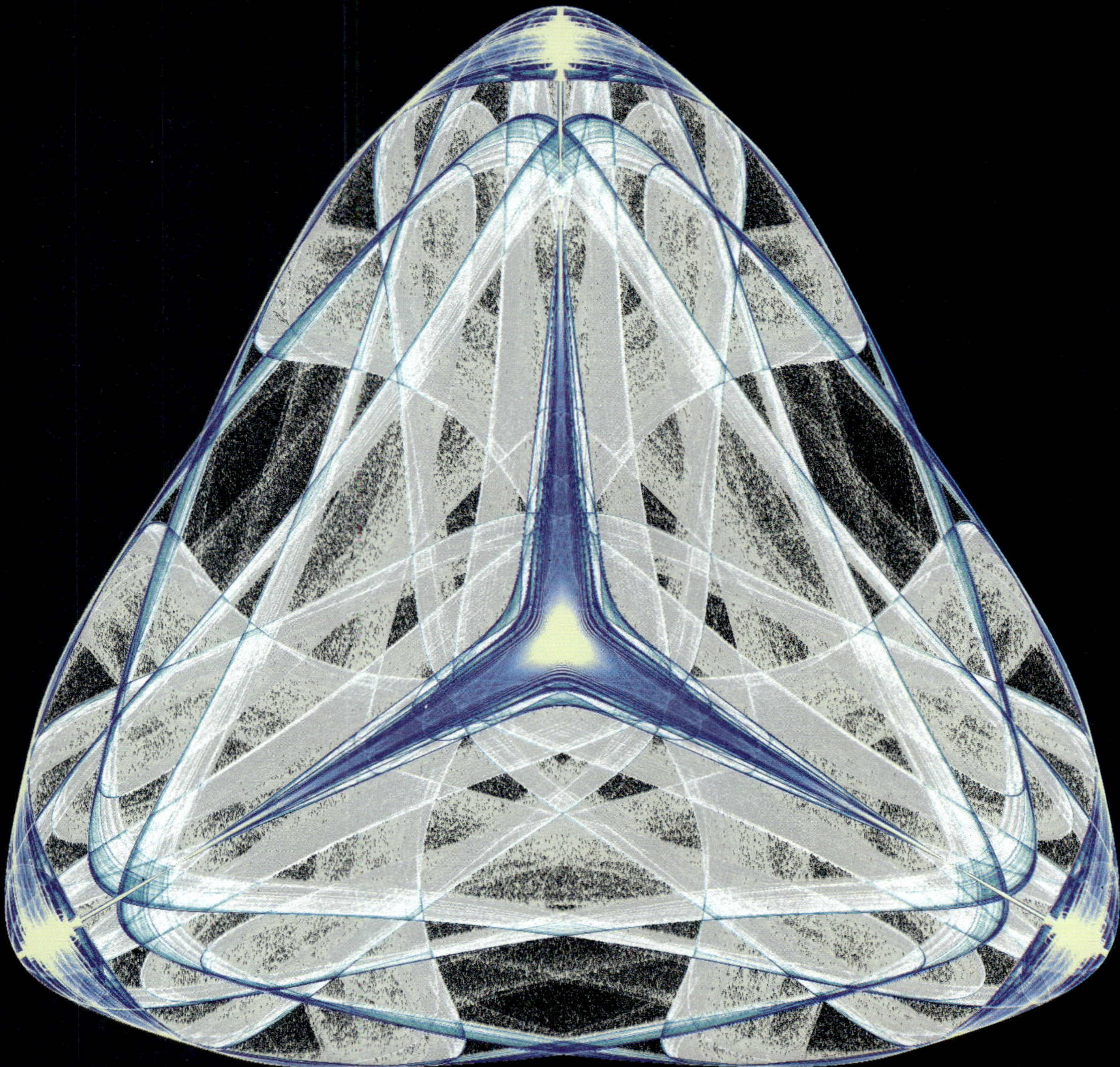

Figure 3.10 Ceramics:
(*a*) (top left) *Ladjvardina ware from fourteenth century Iran,*
(*b*) (top right) *Qianlong moon flask,*
(*c*) (bottom left) *tin-glazed earthenware mosaic from twentieth century Morocco, and*
(*d*) (bottom right) *Sixteenth century Venetian enamelled dish.*

The three-pointed star of Mercedes-Benz is more a symbol of prestige and success, more even than an epitaph to Gottlieb Daimler's dream of conquering travel on land, sea and air. Since this world-famous trademark was first embossed on the radiators of Mercedes cars in the early 1900s (it was not encircled by a ring and proudly raised above the bonnet until 1923) it has stood for the very finest in engineering and craftsmanship.

Once you start looking for symmetric icons you find them everywhere. For example, (see Figure 3.10) in ceramic designs we have the following: a famille rose and doucai Qianlong moon flask with

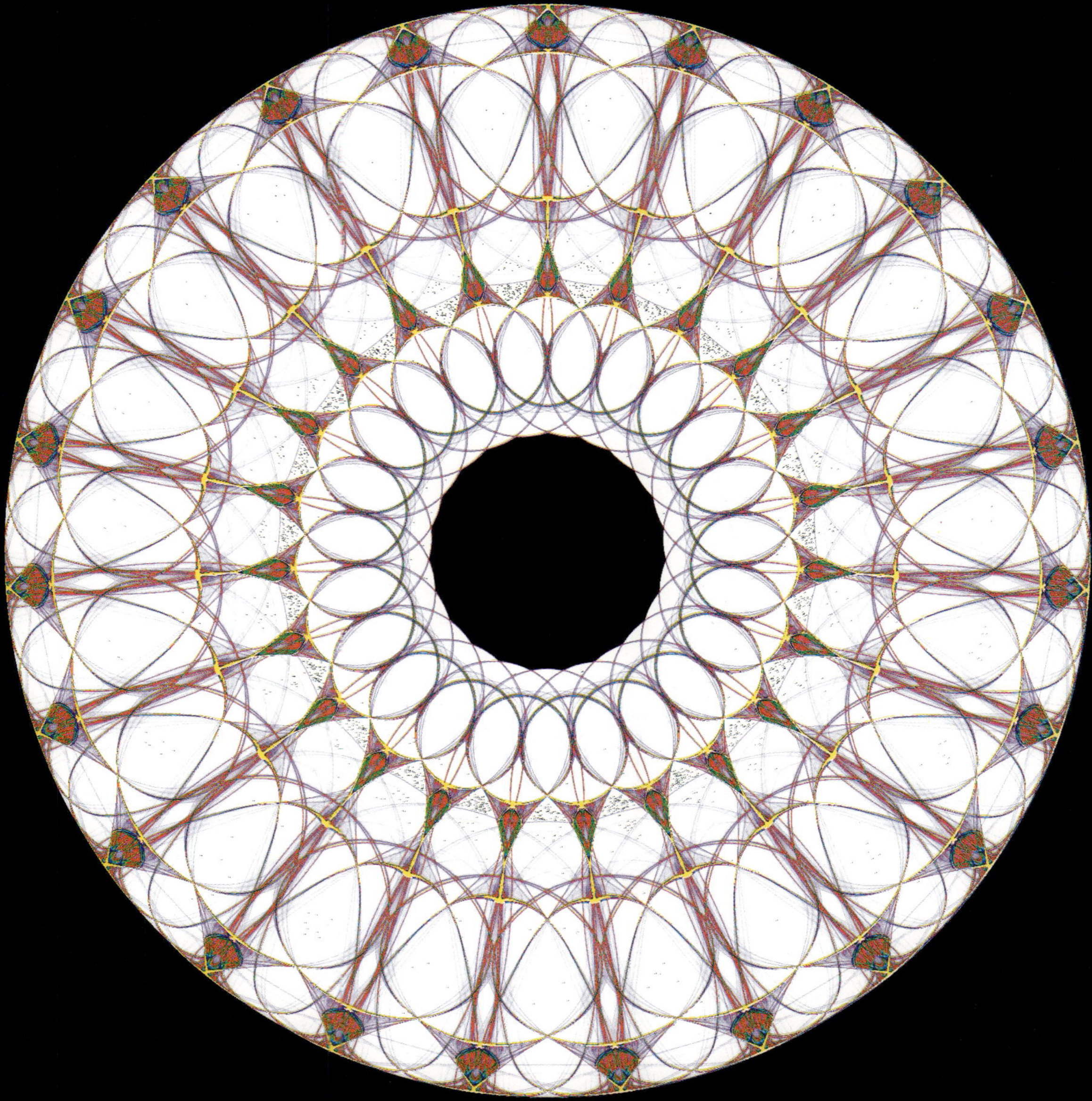

hexagonal symmetry, a sixteenth century Venitian enamelled dish with 42-fold symmetry, Ladjvardina ware from fourteenth century Iran with octagonal symmetry, and a twentieth century tin-glazed earthen-ware mosaic from Fez, Morocco. We contrast these examples of earthenware design with two symmetric icons in Figure 3.11. The 57-fold symmetric icon in Figure 3.11(b) was discovered by Greg Findlow at Sydney.

Figure 3.11 Icons for ceramics: (*a*) (above) *Kachina Dolls. 23-fold symmetry and* (*b*) (over leaf) *Sunflower. 57-fold symmetry.*

Figure 3.12 *Rose window at Chartres*

Some of the most dramatic symmetric icons may be found in the rose windows of Gothic cathedrals. Two particularly beautiful examples are the rose windows at Chartres (Figure 3.12) with twelvefold symmetry that Painton Cowen in *Rose Windows* calls 'a triumph of geometry' and the thirteenth century 24-fold symmetric double layered wheel at Santa Chiara in Assisi Figure 3.13(a).

It is worth quoting Cowen more extensively concerning the meaning of the degree of symmetry in rose windows:

Every rose window is a direct expression of number and geometry—of light in perfect form. At Chartres all of them are divided into twelve segments, the number of perfection, of the universe, and the Logos.

The numbers one to eight were the most important, together with the all-important number twelve, and each had a geometric equivalent. *One* represented the unity of all things, symbolized by the circle and its center; *two*, duality and the paradox of opposites, expressed as pairs across the center; *three*, the triangle, stability transcending duality; *four*, the square, matter, the elements, winds, seasons and directions; *five*, the pentacle, man, magic, and Christ cruci-

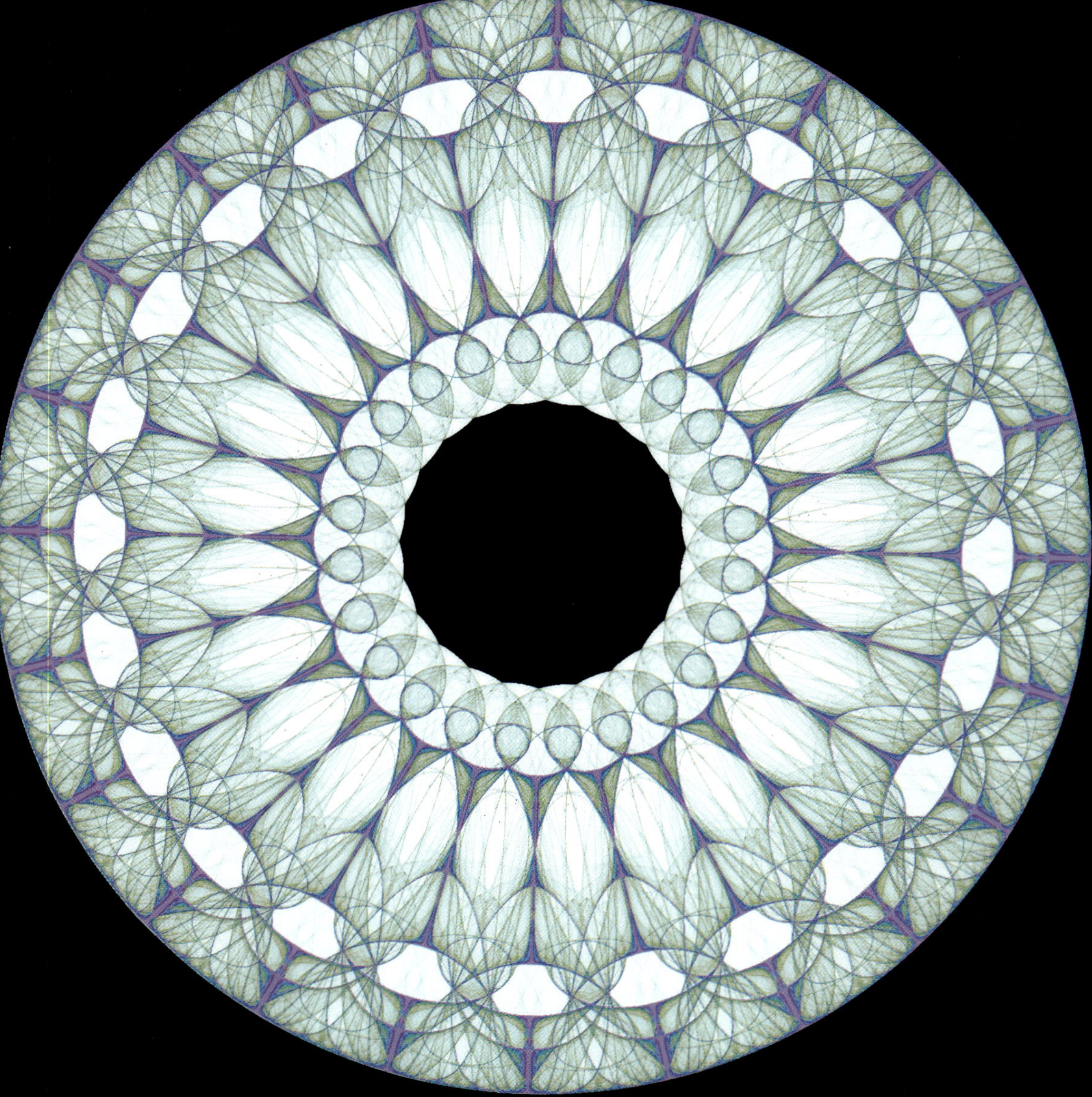

fied with five wounds; *six*, the number of equilibrium and balance within the soul, symbolized in the Star of David or Solomon's Seal; *seven*, the mystic number, of the seven ages, planets, virtues, gifts of the Spirit and liberal arts; *eight*, the number of baptism and rebirth, implied in the octagon; *twelve* that of perfection, the universe, time, the apostles, the Zodiac, the tribes of Israel, and the precious stones in the foundations of the New Jerusalem.

Figure 3.13 (*a*) (opposite below) *double layered wheel at Santa Chiara and* (*b*) (opposite above) *Santa Chiara Icon*

Figure 3.14 (above) *French Glass. A ninefold symmetric icon in the style of stained glass.*

Figure 3.15 *(a) The Pentacle.*

Every rose window contains a star, either placed literally in the rose—as at Saint-Ouen and Sens—or implied through the radiating pattern. In the language of symbolism stars have many meanings, but they primarily relate 'to the struggle of spirit against the forces of darkness'. Thus, the five-pointed pentacle is a symbol of magic, the Pythagorean symbol of healing, of the Crucifixion, and at a later date Man, drawn within the pentacle by Leonardo da Vinci. The six-pointed Solomon's Seal or Star of David is the star of the macrocosm, of heaven

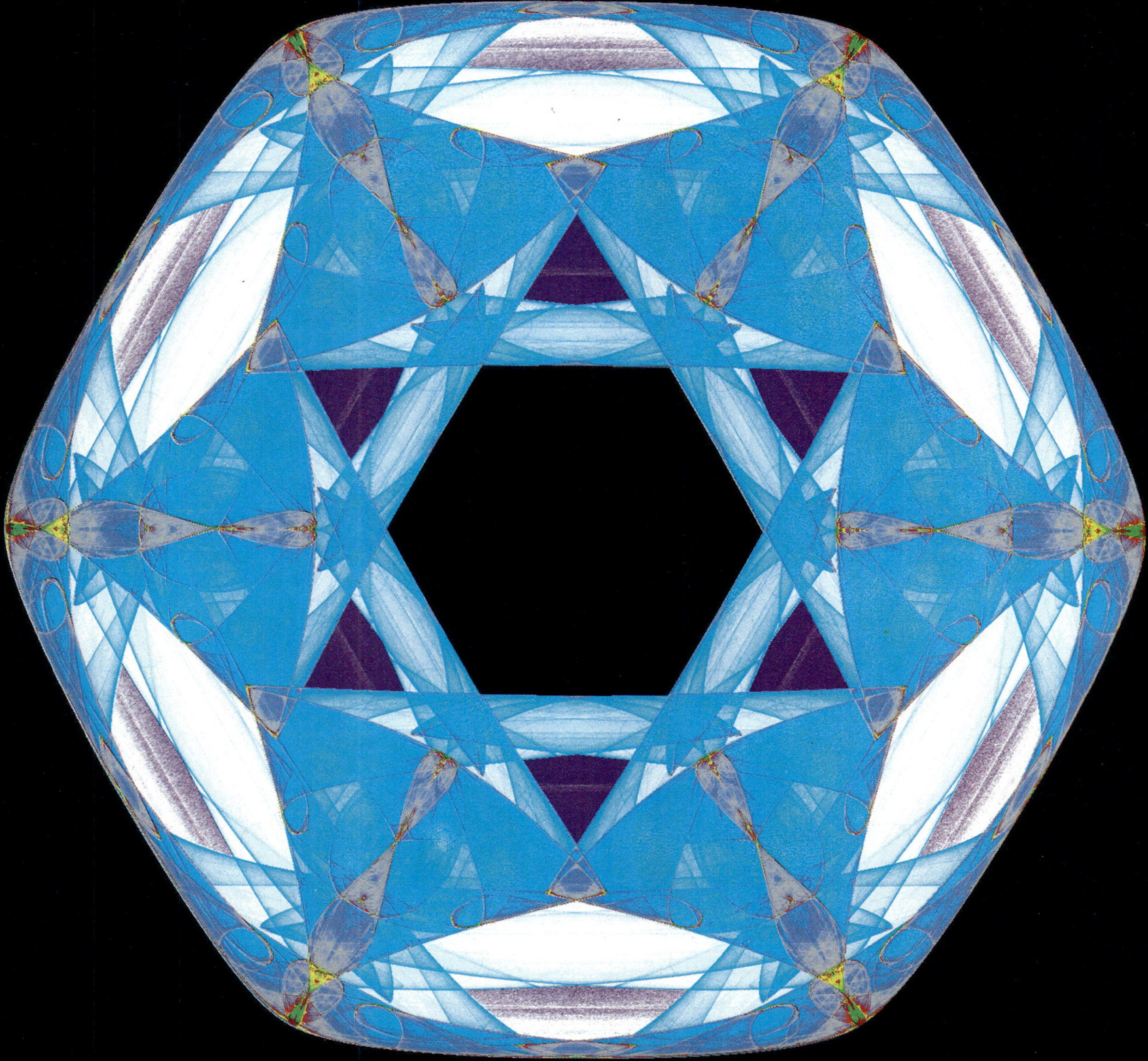

and earth united through man, the two interlocking triangular symbols of fire and water forming the perfect union of the conscious and the unconscious.

In Figures 3.15(a) and (b) we present symmetric chaos icons of the five-pointed pentacle and the six-pointed Star of David. How easily these figures impersonate the semi-mystical icons of religious numerology.

Figure 3.15 *(b) Star of David.*

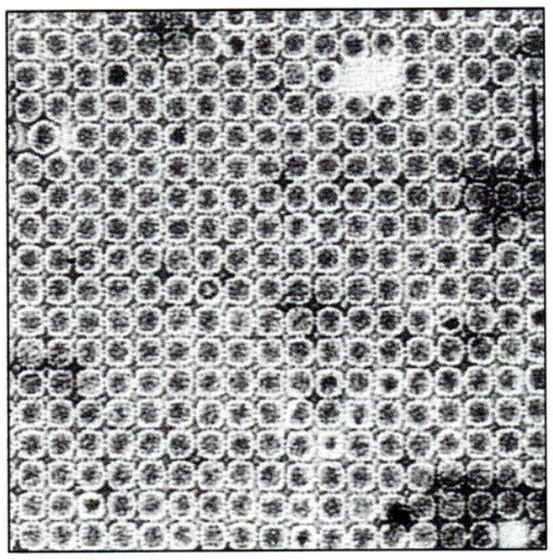

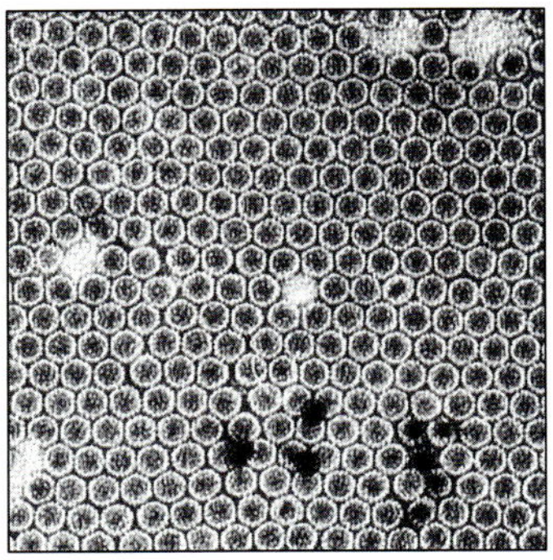

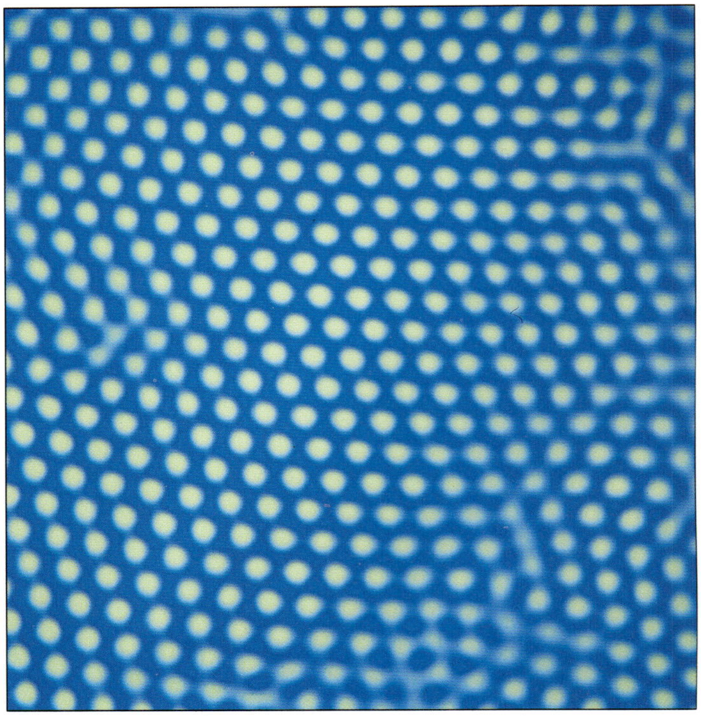

Figure 3.16 *Lattices of plant viruses:* (*a*) (above left) *square and* (*b*) (below left) *hexagonal.*

Figure 3.17 (right) *A stationary chemical pattern.*
Courtesy of Qi Ouyang and Harry L. Swinney.

Tilings

Thus far we have stressed the dihedral and cyclic symmetries of symmetric icons. But there is another class of planar symmetries that are exhibited by quilts and tilings. The main difference between the icons and the quilts is that the icons are bounded but the quilts, through infinite repetition, are not. Some instances of tilings are quite unexpected. For example, a Bromovirus can pack itself in either square or hexagonal arrays (Figure 3.16). The hexagonal pattern in Figure 3.17 illustrates, in an actual experiment, the relative concentrations of certain chemicals.

More familiarly, tilings occur in Islamic art, Italian mosaics (Figure 3.18), quilts themselves (Figure 3.19), Indonesian batiks (Figure 3.21), and ceramics (Figure 3.23).

There are a number of partially preserved examples of Italian mosaic floors that consist of square symmetric patterns on a square lattice. In Figure 3.18(a), we present a Second Century geometric Roman mosaic found in Hatay, South Turkey and along with that a computer generated symmetric chaos design that we call Sicilian Tile (Figure 3.18(b)).

Figure 3.18 (*a*) (opposite below) *Second Century Roman mosaic from Turkey,* (*b*) (opposite above) *Sicilian Tile.*

Figure 3.19 *American quilt designs taken from the catalogue of* The Gazebo of New York.

Figure 3.20 (*a*) (opposite below) *An Indonesian geometric design in stone and* (*b*) (opposite above) *Roses.*

Square tilings are also found throughout Indonesian art. In Figure 3.20(a), we show a stone tiling found in the Tjandi Prambanan on the border between Jogjakarta and Surakarta. Note the strong resemblence with the geometric design in Figure 3.20(b). In Figure 3.21, we picture a modern bạtik featuring a map of Indonesia surrounded by a number of tiling patterns of a classical motif. Note the similarity in design of the chaotic quilt pattern in Figure 3.21(b).

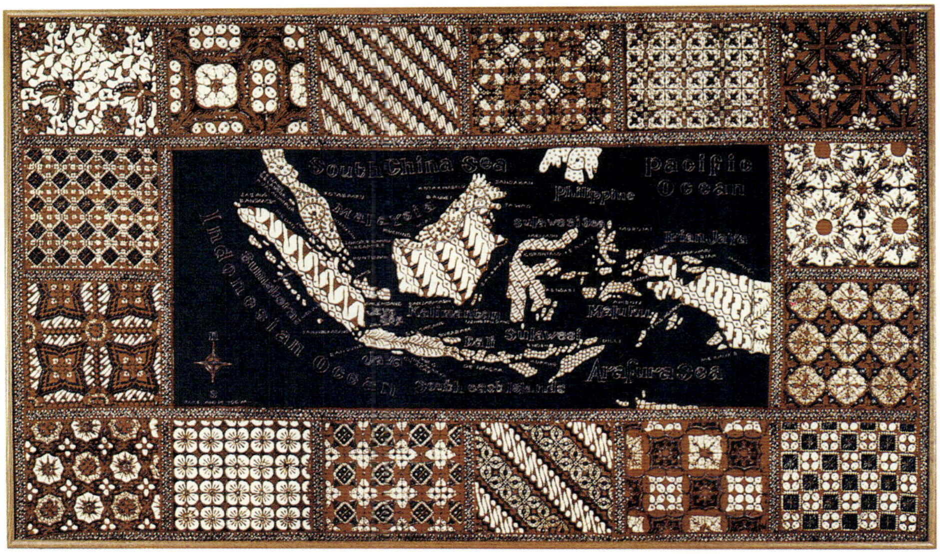

Figure 3.21 (*a*) (left) *A batik map of Indonesian and* (*b*) (opposite) *Wagonwheels.*

Tile patterns can also be found in wrought-iron grills such as the sketch of the fifteenth century choir screen from the Constance Minster shown in Figure 3.22.

Figure 3.22 Fifteenth century choir screen from the Constance minster.

Four samples of Victorian tiles are found in Figure 3.23. Two of these tiles (see Figures 3.23(a) and (b)) have a plain geometric design, and two are rather ornate. The tile in Figure 3.23(c) was produced by Minton, Hollins and Company in the late nineteenth century, while Figure 3.23(d) is unmarked. In Figure 3.23(e) we present our own version of Victorian tiles.

One of the pleasures of experimenting with chaotic quilts on the computer has been the discovery of a seemingly infinite variety of patterns that could be used profitably in wallpaper and ceramic tile design. In Figure 3.24(a) and (b), we present two of these patterns.

Another place where repeating patterns have been utilized with great effect over the years is in the stained glass windows of great cathedrals. In Figure 3.25, we present our own attempt at a stained glass pattern. The square geometry of this design matches well with the strong right angles of the knave windows of Coventry Cathedral.

Figure 3.23 *Victorian ceramic tilings:* (*a*) *and* (*b*) (top left and right) *Unmarked geometric tiles* (*c*) (below left) *Minton, Hollins & Co,* (*d*) *unmarked and* (*e*) (opposite) *Victorian Tiles.*

Figure 3.26 *Arabian hexagonal tiling reproduced from the Dover edition of Owen Jones,* Grammar of Ornament, *originally published in 1856.*

Islamic art

Repeated patterns and symmetry enjoy a special place in Islamic art. In part, this is because of the avoidance by the Islamic faith of art that imitates nature. In practice, Islamic art often uses an ornate decoration, based on principles of symmetry, interwoven with a highly developed calligraphy.

Symmetric tilings of one form or another have been frequently used as decoration on buildings in the Islamic world. In Figure 3.26, we show a tiling based on a hexagonal lattice with hexagonal symmetry recorded by Owen Jones during his travels to the Middle East between 1841–45. In 1856 he published a book *The Grammar of Ornament*

Figure 3.24 (on prevous pages) *Chaotic wallpaper and ceramic designs:* (*a*) (page 90) *Mosque and* (*b*) (page 91) *Red Tiles.*

Figure 3.25 (*a*) (opposite below) *knave windows at Coventry Cathedral* (*b*) (opposite above) *Cathedral Attractor.*

Figure 3.27 (*a*) (above left) *A cupola from La Kleba Mosque at Cordova and* (*b*) (above right) *a vestibule ceiling the Madresseh Chenar Bagh at Ishfahan*

Figure 3.28 *Hexagonal lattice patterns:* (*a*) (right) *Frontispiece of Qur'ān,*
(*b*) (opposite) *Hexagonal Design. symmetric chaos.*

which was one of the first and arguably the most influential design source books on the subject. In Figure 2.15, we show a portion of the ceiling decoration from the Dome of the Rock in Jeruselem. This provides a beautiful example of interlace patterns on a square tiling, in Islamic art. In Figure 3.27 we show cupolas over La Kleba at Cordova, which has eightfold symmetry, and over the Madresseh Chenar Bagh at Isfahan, which has 16-fold symmetry.

Some of the most beautiful examples of the use of symmetric patterns are to be found on decorative pages from books, notably the Qur'ān. In Figure 3.28(a), we show the frontispiece for the Qur'ān, drawn in 1310 by Alī ibn Muhammad ibn Zayd al-'Alavī al-Husaynī in Mosul. We complement this picture with a chaotic design created on a hexagonal lattice Figure 3.28(b).

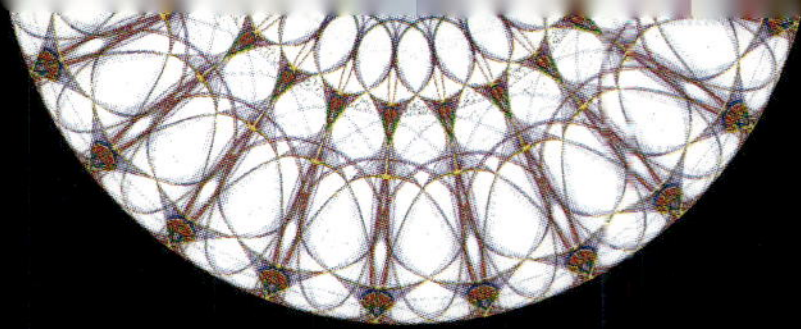

Chapter four

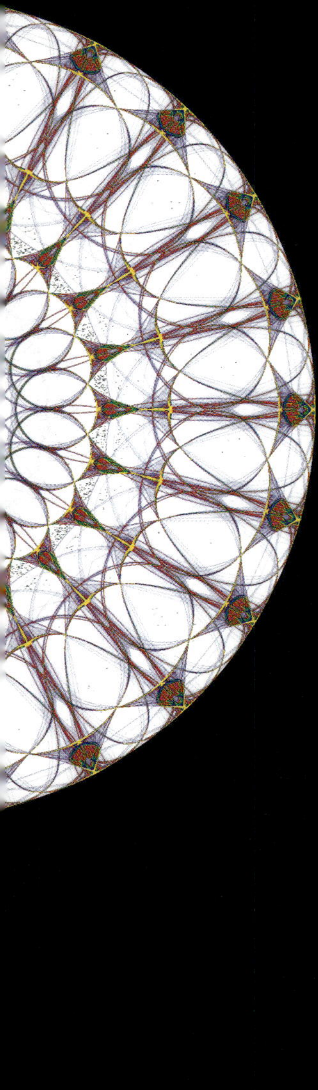

CHAOS AND SYMMETRY CREATION

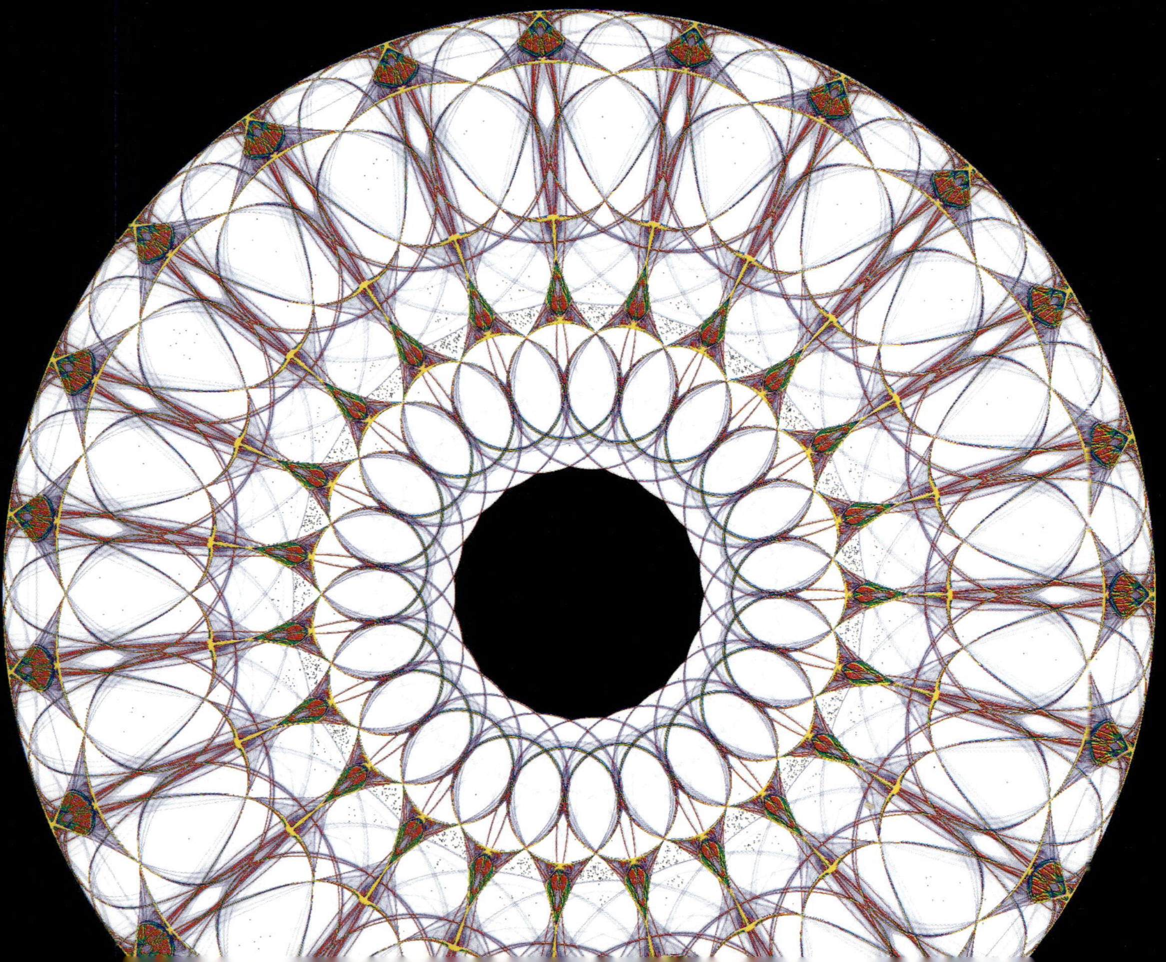

ABOVE all, the power and utility of mathematics is seen in its use for modelling natural phenomena. This is as true now as it was when Newton invented and used the calculus for his law of gravitation. There are two fundamental, and often conflicting, issues in the use of mathematical models. In the first place, one attempts to derive a complete model that can lead to accurate predictions. The model we have of the solar system, based on Newton's laws, is of this type. However, it is usually not enough just to be able to make accurate predictions. In most cases, we need a good understanding of the phenomena modelled by the mathematics. We need this understanding for the control and description of the processes we are modelling. This is seen, for example, in economics where, in spite of highly elaborate models, it is often a matter of great controversy as to exactly what one should do to obtain a desired outcome. The situation is further complicated by the fact that in many situations the available mathematical models are limited by either insufficient knowledge or insufficient data. The study of the climate and economics suffer from both these limitations. Even worse, if one has a complete mathematical model, it may turn out to be largely mathematically intractable. The models we have for turbulence tend to fall into this category.

What are we to do in this situation? Of course, we try to develop models that give accurate predictions. Even though there are many problems concerning the solar system and gravitation that we cannot answer, there is no doubt that Newton's laws have proved of enormous value in the prediction of tides and eclipses and the computation of the orbits of satellites. In some instances, however, there is just no hope of making truly accurate models and one is forced to rely on the study of simple models in order to identify and understand important phenomena. One such area is population dynamics, where long-range prediction has proved elusive, but simplified models have proved helpful. A well-known example of a population model is the logistic mapping, which provides insight into both why the dynamics of populations are so difficult to understand and why chaotic dynamics are so simple to generate.

We shall start our discussion of chaos with a description of the logistic map and how it arises as a model in the study of population dynamics. At the end of the chapter, we show how a mathematical variant of the logistic map, the odd logistic map, illustrates the phenomenon of *symmetry creation* that lies behind the formation of symmetric icons.

The logistic map

The study of chaotic dynamics was made popular by the computer experiments of Robert May and Mitchell Feigenbaum on a mapping known as the *logistic map*. The remarkable feature of the logistic map is the contrast between the simplicity of its form (it is a polynomial mapping of degree two) and the complexity of its dynamics.

The logistic map is the simplest model in population theory that incorporates the effects of both birth and death rates. We imagine an experiment where a census is taken each year of the population of rabbits in an isolated region in South Australia. The basic question facing population dynamicists is to find a model that accurately predicts the future rabbit population, given that the present rabbit population is known. This innocent seeming question is much too difficult to answer with present knowledge. A more modest approach is to construct a model of the rabbit population that gives some insights into what may actually happen. In other words, we try to create models that give insight into how population dynamics works. It turns out that the logistic map is the simplest model that we can construct that incorporates a realistic mechanism for birth and death. In spite of the rudimentary character of the model, the logistic map displays an astonishing range of complex dynamical behavior. Indeed, the simplicity of this model provides compelling evidence that the real dynamics of rabbit populations, whatever they are, are likely to be at least as complex as those of the logistic map.

At this stage, it is helpful to introduce some more economical notation to describe the future population of rabbits. To this end, we shall let p_n denote the population of rabbits after n years and, in particular, let p_0 denote the present rabbit population.

We begin by deciding what factors might influence p_n. We shall start by making the reasonable assumption that the population in any given year depends only on the previous year's population: in symbols this means that there is a rule g such that

$$p_{n+1} = g(p_n).$$

Otherwise said, we have a formula 'g' that gives the population of rabbits after $n - 1$ years in terms of the population after n years. Next we discuss what form the rule g should have. A natural assumption to make is that the rule g depends only on the birth and death rates of the rabbit population. Moreover, we shall assume that the birth and death rates do not depend on the year (though we shall allow the possibility that they depend on the *size* of the population). Finally, in order to determine g, we must specify how the birth and death rates actually enter the formula for g.

If there are no deaths and the birth rate is constant and

equal to b, then the simplest dependence of the rabbit population on the birth rate would be

$$p_{n+1} = p_n + bp_n.$$

In words, this formula states that the rabbit population after $n + 1$ years is the sum of the rabbit population after n years (that is, p_n) with the number of new rabbits born in year n (that is, bp_n). For example, if $b = 1$ (a rather low birth rate for rabbits), we would have

$$p_{n+1} = p_n + p_n = 2p_n,$$

and so the model for the population of rabbits would obey the doubling rule of Chapter 1. In any case, whenever the birth rate is greater than zero, it is easy to see that in this model the population grows without bound, leading to the absurd conclusion that after a few years the rabbit population would cover the land of South Australia to a depth of several hundred feet.

In short, any reasonable model must incorporate a mechanism for death, presuming that the population of rabbits is to remain bounded. One way of incorporating death into our model is to assume that a constant proportion of the population dies each year and that new baby rabbits are born only to that portion of the population that survives. For example, if the population at the start of the year were N rabbits, we might assume that over the year only sN rabbits reproduce and the rest die (think of s as the *survival* rate). Consequently, at the end of the year there would be $(1 + b)sN$ rabbits leading to the model

$$p_{n+1} = (1 + b)sp_n.$$

Unfortunately, this model has the same difficulty as the previous one. If the *growth rate* $(1 + b)s$ exceeds unity, then the population still grows without bound. There is a new feature that appears in this model. If the survival rate is so small that the growth rate is less than unity, then the model predicts that eventually all of the rabbits will die (this should be compared with the halving map of Chapter 1). Of course, in the unlikely event that the growth rate is exactly 1, the population will remain constant for all time. Since rabbits have a way of surviving, even though their numbers do not grow without bound, our model is unrealistic and cannot be correct.

From a mathematical point of view, the difficulty with both of the previous models is that they are *linear*: next year's population is just a *constant* times this year's population (though the exact interpretation of the constant differs in the two models). Linear models lead naturally either to exponential growth or to exponential extinction, neither of which is a satisfactory conclusion. It follows that a model must be nonlinear if it is to yield the realistic predictions of bounded and positive population size.

From a modelling point of view, one way to address the difficulty of unbounded population growth is to assume that there is a maximum number of rabbits P that can be supported on the land, where the constant P depends on such features as the availability of food and the existence of predators. In such a model, it is reasonable that the survival rate is proportional to the proximity of the population to P. In other words, we might suppose that if the population at the beginning of the year was N, then the survival rate would be $s'(P - N)$ and so $s'(P - N)N$ rabbits would survive to the end of the year and reproduce. We assume here that s' is a constant which does not depend on the population size (or year). This reasoning leads to the new model

$$p_{n+1} = (1 + b)[s'(P - p_n)]p_n.$$

Finally, we can simplify this formula by setting $x_n = p_n/P$, the proportion of the maximum population present at year n, and $\lambda = (1 + b)s'P$. The number x_n is called the relative population; it is the ratio of the actual population to the maximum possible population and is therefore a number between 0 and 1. In terms of x_n, this model simplifies to

$$x_{n+1} = g(x_n) \equiv \lambda x_n(1 - x_n),$$

where g is called the *logistic* mapping and λ is called the *effective growth rate*. Moreover, there is a natural limit on the size of λ in this model. Since we obtain each year's relative population by applying the mapping g to the previous year's relative population, we must assume the result of applying g to a number between 0 and 1 is also a number between 0 and 1. Thus, in order for this formula to make sense as a model of population growth, the function g can never achieve a value greater than unity when $0 < x_n \leq 1$. Since the maximum value of g is $g(\frac{1}{2}) = \frac{1}{4}\lambda$, we must assume that $0 \leq \lambda \leq 4$. Also note that, as promised, the mapping g is nonlinear—there is a quadratic term x_n^2 in its definition. Moreover, the mapping g has the simplest form possible for a nonlinear mapping.

The period-doubling cascade

What May and Feigenbaum each explored is how the dynamics of the logistic mapping g changes as the effective growth rate λ is varied. We discuss here some of their results. Suppose that $\lambda < 1$. Then, since $1 - x_n$ is always less than or equal to 1, we see that

$$x_{n+1} = \lambda x_n(1 - x_n) \leq \lambda x_n < x_n.$$

Hence the population always decreases in size and, as with our earlier population model, the population decreases to zero as n increases and the rabbits disappear.

Next, suppose that the effective growth rate λ is greater than 1. Then there is a critical relative population x^* that remains constant year after year. More precisely, the result of applying g to the point x^* is just x^*. Thus, should x_n ever attain that critical value of x^* the model g predicts that the relative population will maintain its size each year after the nth year. The value x^* is called a *fixed point* for the map g. It is relatively easy to verify that the logistic map has precisely one nonzero fixed point $x^* = 1 - 1/\lambda$. Moreover, when λ lies between 1 and 3 this fixed point is attracting in the sense that as we increase n, the population will always approach x^* (unless we started with x_0 equal to zero or one!). In this sense, the map mimics the halving map considered in Chapter 1, which had an attracting fixed point at zero.

Thus far, we have seen that when $\lambda < 1$ the population becomes extinct, while if λ lies between 1 and 3 the population stabilizes over time to x^*. It is easy to conceive of other possibilities. For example, suppose that we can find a point x_1^* such that

$$x_2^* = g(x_1^*) \neq x_1^*$$

(x_1^* is not a fixed point of g) and

$$g(x_2^*) = x_1^*.$$

If the initial population is x_1^*, then that relative population repeats every *second* year. We call x_1^* a point of *period two* for g. We note that x_2^* is also of period two, and indeed points of period two always come in matched pairs. In fact, the idea that the population can alternate in size from year to year is quite reasonable. If we start with a large population, then resources will be scarce and we can expect a large number of deaths leading to a small population. On the other hand, if we have a small population, resources are plentiful and we can expect an explosive growth in the population size. Given the right conditions, it is reasonable to presume that the relative population will continue cycling between large and small populations *ad infinitum*. With these thoughts in mind, let us return to the study of what happens in the logistic map as we continue to increase λ.

We find that as λ increases through 3, a pair of period-two points is created. Moreover, for almost all initial populations of rabbits (that is, values of $x > 0$), the population x_n approaches the pair of period-two points as the years pass. In alternate years the population of rabbits is high and then low. This change in dynamic behavior is called a *period-doubling* bifurcation.

The period-doubling bifurcation is intriguing, but pales in significance before the period-doubling *cascade* which follows. As λ is further increased, the period-two point itself period-doubles to a period-four point (the population of rabbits repeats every four years); then period-doubling to period eight occurs, then to period sixteen, and so on. It is astonishing that a model as elementary as the logistic map actually predicts the possibility of complicated periodic dynamics. From this point of view, the 14-year cycles of locusts should perhaps not be considered such a remarkable phenomenon.

When λ reaches $\lambda_c = 3.53\ldots$, periodic points of arbitrarily large period have appeared in the dynamics of g and chaotic dynamics has set in. This bifurcation scenario is summarized by the now famous bifurcation diagram pictured in Figure 4.1. In this figure the horizontal axis shows the parameter λ and the vertical axis shows the population x. For each value of the parameter, the attracting set—first the fixed point, then the period-two points, then the period-four points, etc.—are shown.

We now describe how Figure 4.1 is produced. We will need this information when we describe a similar bifurcation diagram for symmetric mappings. There are five steps in the procedure, which we now enumerate.

1. Choose a value of $\lambda > 1$ and a value of x between 0 and 1.
2. Using the logistic mapping g iterate 100 times starting from x to compute the transient part of the dynamics.

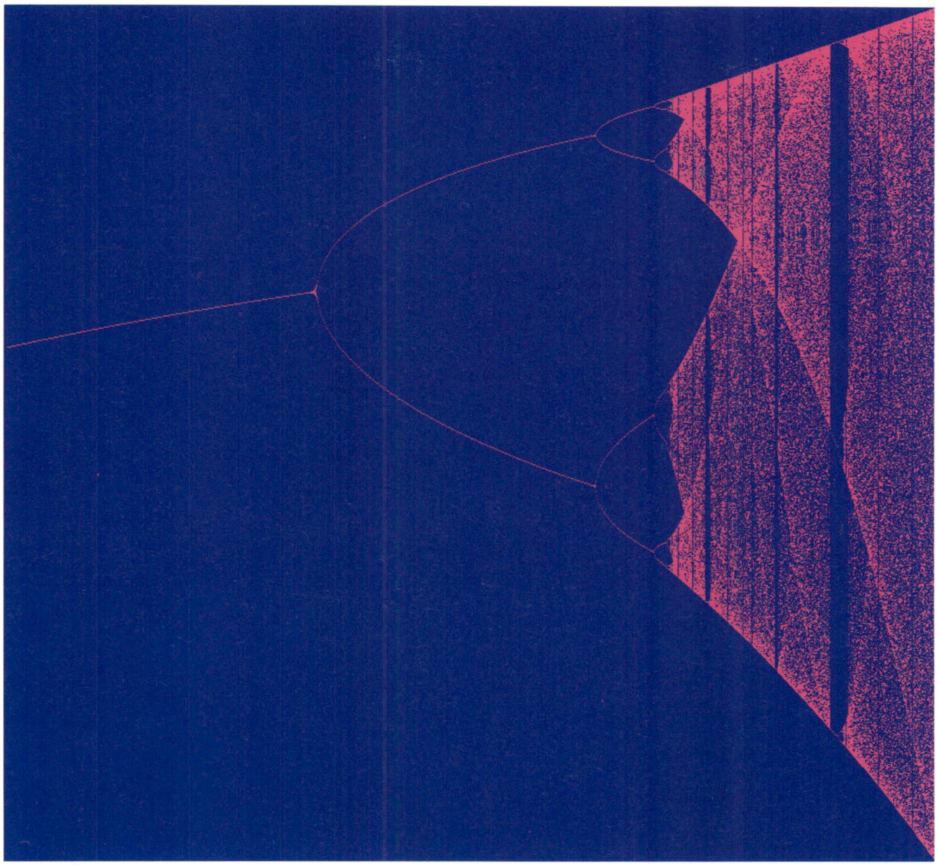

Figure 4.1 *The period-doubling cascade in the logistic mapping.*

3. Plot the next 200 iterates of x on the vertical line above the current value of λ on the λ-axis.
4. Increment the value of λ by some small amount, say 0.05.
5. Return to step 2 using the current value of x.

In this way we see the fixed point turning into a period-two point as λ is increased, and then into a period-four point, etc.

There are two issues that are important here. The first is the discovery of period-doubling and the period-doubling cascade in this simple logistic map. The consequence of having this cascade is that small changes in a single parameter, such as the effective growth λ, can create large changes in the observed dynamics. The second is the realization that the kind of complicated dynamics that is observed in the logistic equation is typical of the dynamics for a large class of mappings.

It was a major surprise to most researchers when in 1976 Robert May pointed out that even mathematical models so simple as the logistic map can predict such complicated—or chaotic—dynamics. It has taken time for people to realize that chaotic dynamics might even be more the rule than the exception.

In 1978 Mitchell Feigenbaum made another remarkable discovery about the logistic map and mappings like the logistic map: there is a universal constant of mathematics (akin to π and e) that can be associ-ated with period doubling cascades and, moreover, this constant can even be determined experimentally. Indeed, this discovery was so surprising that initially there was great reluctance in accepting Feigenbaum's ideas.

Feigenbaum's contribution was the observation that to each period-doubling cascade one can compute a number as follows. Suppose that a period doubling cascade is found by varying a parameter λ, like the λ in the logistic map. Suppose that you let λ_m be the parameter value where the mth period-doubling occurs. Then the ratio

$$\frac{\lambda_m - \lambda_{m-1}}{\lambda_{m+1} - \lambda_m}$$

tends to a constant as m becomes large and this constant is now known as *Feigenbaum's number*. What is important is that there is a large class of mappings, which includes the logistic map, for which this number is the same. That number is 4.6692016. . . .

Symmetric maps on the line

As we mentioned at the beginning of this chapter, one of the two roles of mathematical modelling is to identify and illustrate new phenomena. From this point of view, the study of the logistic map has been a glorious success. The study of this basic mapping has led to a better understanding of period-doubling and the role of the period-doubling cascade in leading to the existence of chaotic dynamics. It has also demonstrated why we should not be surprised to find that the dynamics of real populations are extraordinarily complicated.

Suppose now that one wants to study how symmetry and chaotic dynamics coexist. Since the logistic map has a particularly simple form and also exhibits chaotic dynamics, it makes sense to try to modify the logistic map so that it is also symmetric. We begin this process of modification by discussing symmetry on the line.

Note that the real line has only one nontrivial symmetry that fixes the origin, namely, reflection about the origin:

$$Rx = -x.$$

What does it mean for a *map* to be symmetric with respect to R? Suppose that F is a mapping of the line. That is, for each number x, F determines a new number that we denote by $F(x)$. In order that F be symmetric, we shall require that if we evaluate F at symmetrically placed points x and Rx, we obtain a pair of *symmetrically placed* points $F(x)$ and $F(Rx)$. That is, we require

$$F(Rx) = RF(x).$$

Since Rx = −x, we see that this condition on F amounts to requiring

$$F(-x) = -F(x)$$

In other words, F is R-symmetric if it is an *odd* function. It is easy to check that a polynomial $F(x)$ is odd precisely when the monomials that appear in the formula for F are of odd degree. For example, if $F(x) = x + x^2$ then F is odd, but if $F(x) = x + x^3$ then F is not odd. In particular, the logistic map

$$g(x) = \lambda x - \lambda x^2$$

is not an odd function, on account of the x^2 term.

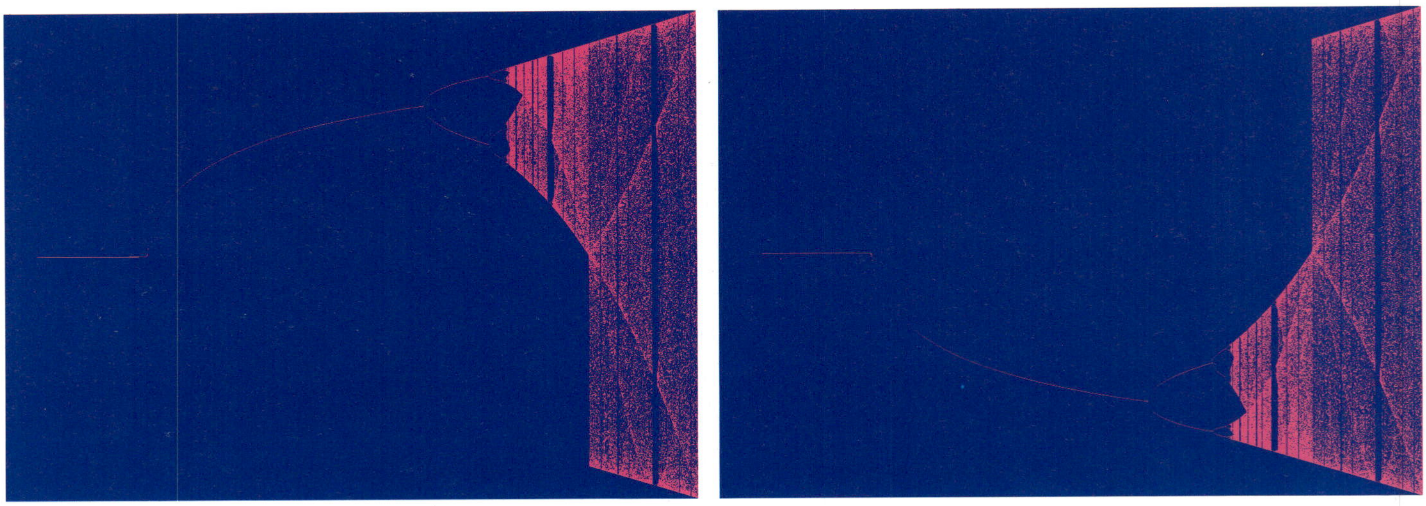

(a) (b)

Figure 4.2 *Symmetry creation in the odd logistic map.*
(a) positive initial condition and
(b) negative initial condition.

Next, we modify the logistic map so that it is an odd function. What we do is define

$$g_o(x) \equiv \lambda x(1 - x^2) = \lambda x - \lambda x^3.$$

We call g_o the *odd logistic map*. Unlike the logistic map, our new map has no obvious interpretation as a model for population dynamics. Nevertheless, we can hope that if there are new phenomena that occur in the dynamics of symmetric mappings, then these phenomena will become visible when we study the dynamics of the odd logistic map.

Symmetry creation

The first step in studying the dynamics of the odd logistic map is to create by computer its bifurcation diagram. What we see is that the bifurcation diagram of the odd logistic map has many features in

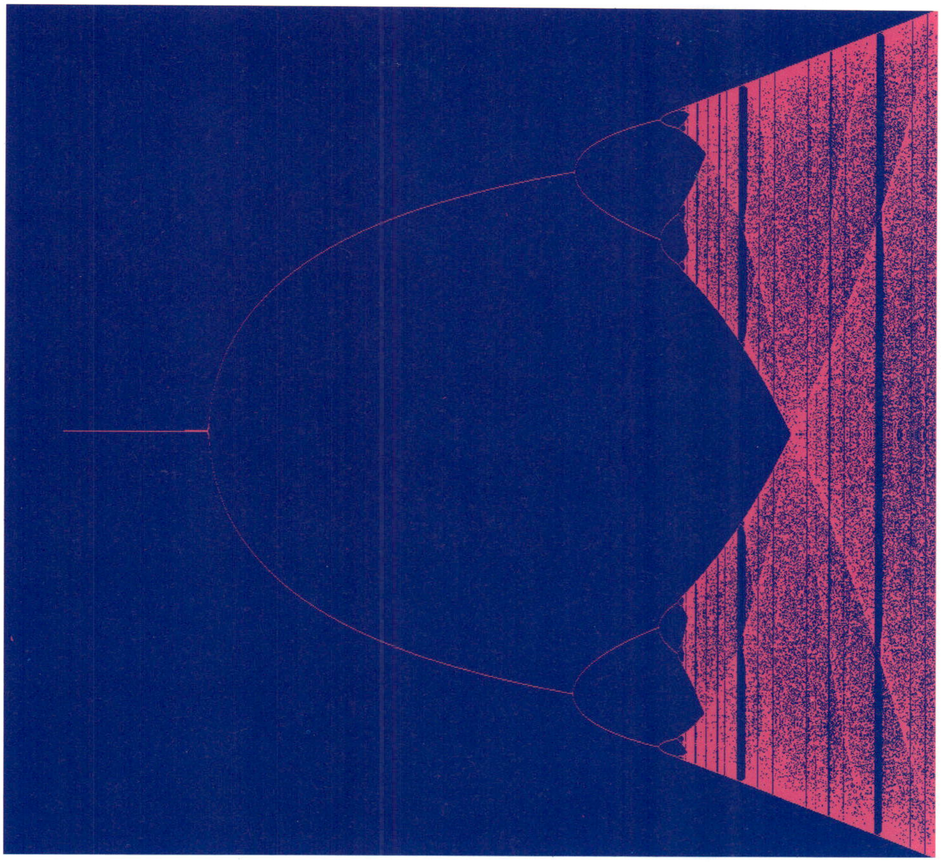

Figure 4.3 Conjugate attractors in the odd logistic map.

common with the bifurcation diagram of the logistic map. In particular, we find that the odd logistic map has a period-doubling cascade. However, because of symmetry it is forced to have two cascades: one corresponding to choosing a positive number x as the initial point in the iteration procedure, as in Figure 4.2(a), and the other corresponding to choosing a negative number x as initial point, as in Figure 4.2(b). We show in Figure 4.3 the superposition of the two bifurcation diagrams from Figure 4.2

As we increase λ, we find that the two chaotic attractors spawned by the period-doubling cascades merge to form one attractor with greater symmetry. (The value of λ where this merging of attractors occurs is at $\lambda_c = \frac{3}{2}\sqrt{3}$.) Below λ_c initial conditions lead to attractors all of whose points are positive or to attractors all of whose points are negative. Above λ_c the chaotic attractors are symmetric about the origin. In Figure 4.2, the bifurcation diagram showing this symmetry creation for the odd logistic equation is presented; this diagram is produced in the same fashion as was Figure 4.1, but with the logistic map replaced by the odd logistic map. The odd logistic equation provides the most elementary example of a symmetry increasing bifurcation.

At the end of Chapter 1, we indicated how the kind of symmetry creation exhibited by the odd logistic map might lead to *patterns on average*. We now present another example of symmetry creation.

Chaotic trains

For a number of years, engineers have known that the wheelsets of trains may move chaotically. These wheelsets are designed to allow some lateral motion. Indeed, this sideways motion is clearly needed on curves, but it is also needed on straight stretches, as tracks are not always perfectly aligned. Chaotic motion of the wheelsets is known to occur and this chaotic motion does not bode well for passenger comfort and wheelset durability.

Recently, in a paper to appear in the *Philosophical Transactions of the Royal Society of London*, three Danish engineers—Carsten Knudsen, Rasmas Feldberg, and Hans True—have studied a detailed model of the lateral motion of wheelsets and have found an instance of symmetry creation. Like all models of wheelsets, their model has a single left–right symmetry and it is this symmetry that is involved in the symmetry creation.

What these engineers found is that at slow speeds of about 25 miles per hour (mph) the sideways motion can be chaotic—but asymmetric. More precisely, the lateral position of the wheelset (relative to the track) is biased either towards the right or towards the left. At such speeds small imperfections in the track have the tendency to force the train to position its wheels in one preferred direction. This preference can cause one of the wheels, say the left, to wear down more quickly than the right; the difference in the diameters of the wheels then forces the wheelset to turn towards the side with the preferential wear, in this case the left, which results in even greater differential wear. This is clearly an unstable and unwanted situation, and explains why railroad engineers might be interested in asymmetric chaotic motion.

At higher speeds of about 35 or 40 mph, the model predicts a transition to symmetric chaotic motion in a way that very much resembles the transition from asymmetric to symmetric chaos in the odd logistic map. What is curious about this discovery is the possibility that wheelset repairs might be more frequent when trains travel at *slower* speeds. Of course, once such a phenomenon is identified, wheelset designs can be adapted to guard against such differential wear. Nevertheless, it is intriguing that the phenomenon of symmetry creation can appear as a significant factor in practical engineering problems.

Chapter five

SYMMETRIC ICONS

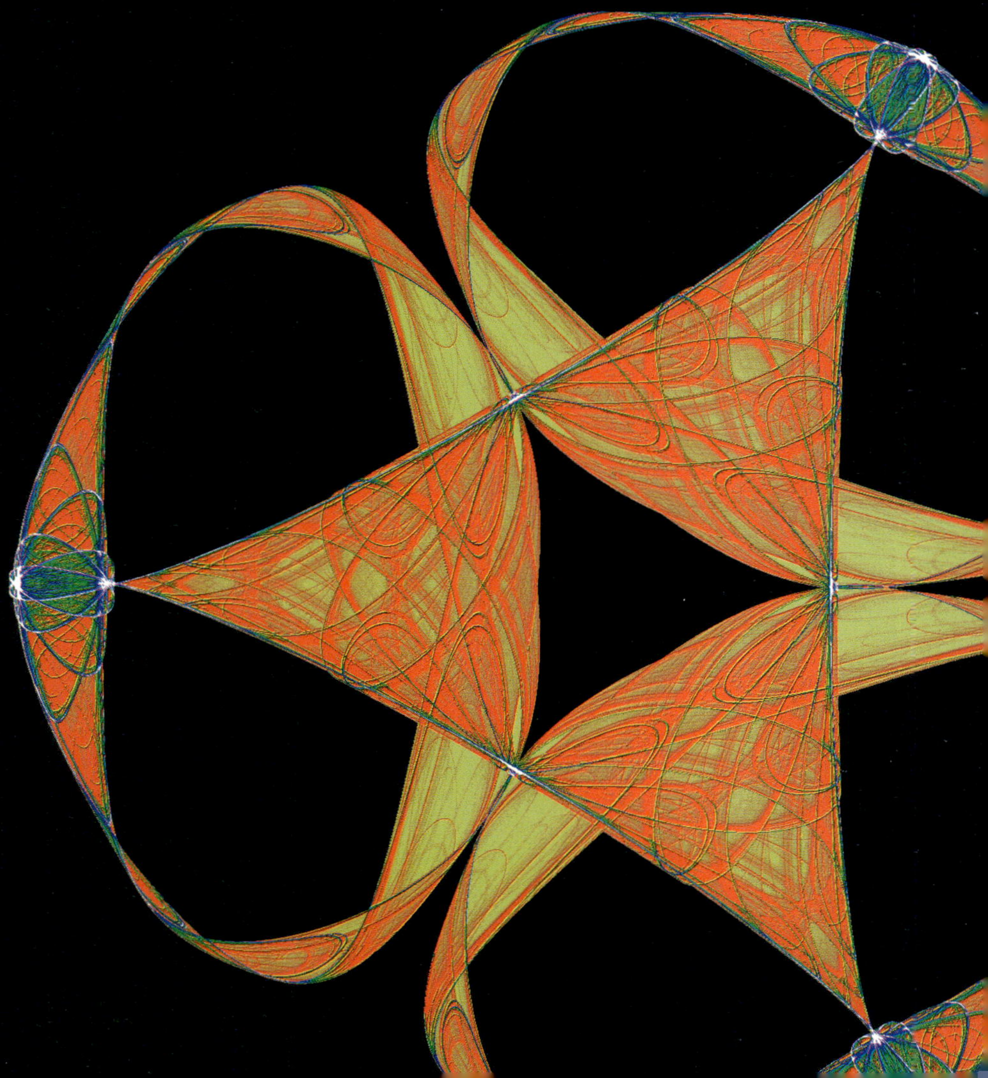

In this chapter, we focus our attention on the properties and construction of symmetric icons. As a first step, we provide a striking pictorial demonstration that the mathematical rules we use to construct symmetric icons possess *sensitive dependence on initial conditions*—one of the characteristic properties of chaos. Next we give more details, for the mathematically inclined, on the explicit mathematical formulas that we have used throughout the book to construct icons in the plane with $\mathbf{D}_n$ symmetry. These formulas can be viewed as a two-dimensional symmetric version of the logistic map and are most easily described using complex numbers. For this reason, we include a brief introduction to the history and properties of complex numbers. We conclude the chapter with a demonstration of symmetry creation for mappings of the plane which possess $\mathbf{D}_n$ symmetry.

Mappings with dihedral symmetry

We begin by discussing how we might extend the idea of symmetry on which the odd logistic map is based to maps of the plane with dihedral symmetry. This extension will eventually provide us with the formulas that we have used to create all of the symmetric icons shown in this volume.

Let us start by supposing that we are given a planar mapping F. That is, for each point z in the plane, F determines a new point in the plane that we denote by $F(z)$. Guided by our discussion of the odd logistic map, we may answer the question as to what it means for F to be $\mathbf{D}_n$ symmetric. Suppose that S is a symmetry in $\mathbf{D}_n$. We require that the symmetrically placed points z and Sz are taken by F to symmetrically placed points. More precisely, we require that

$$F(Sz) = SF(z)$$

for each symmetry S in $\mathbf{D}_n$. When F is $\mathbf{D}_n$ symmetric, this formula allows us to compute $F(Sz)$ once we know $F(z)$.

Subsequently, we shall give explicit formulas for symmetric maps. However, for a little longer, we shall keep the formulas hidden and instead give a graphic illustration of how we can exploit the symmetry of a map to display the hallmark of chaos—sensitive dependence on initial conditions.

An illustration of chaos

As we discussed in the introductory chapter, one of the most important characteristics of chaotic dynamics is sensitive dependence on initial conditions. Let us recall what this means in the context of symmetric mappings of the plane. Suppose that when we begin iterating a point z sufficiently near the origin, all iterates of z stay within a bounded region of the plane. (This assumption is valid for the mappings used to create the symmetric icons pictured in this volume.) Choose two initial points z_1 and z_2 that are very close together. We say that our mapping has sensitive dependence on initial conditions if, when we compute the sequence of iterates for both z_1 and z_2, the corresponding points in the sequence move away from each other. For example, if z_1 and z_2 lie in the same pixel of the computer screen, we would require that corresponding iterates of z_1 and z_2 eventually lie in different pixels (see Chapter 1). In fact, this property cannot be expected to be valid for every pair of points in a pixel, but when sensitive dependence is present we expect it to be valid for almost every choice of points.

Figure 5.1 *Color coding of wedges between lines of symmetry.*

Figure 5.2 *Color coding of the first iterate.*

Figure 5.3 *Color coding of the fourth iterate.*

Coupling symmetry and color allows us to illustrate pictorially the notion of sensitive dependence. To understand how this is possible, we restrict our attention to mappings with triangular symmetry. Suppose that f is a mapping of the plane with $\mathbf{D}_3$ symmetry. Then any point of the plane which lies on an axis of symmetry of $\mathbf{D}_3$ must be mapped onto that same axis of symmetry. To see why this is so, we start by observing that the three axes of symmetry of $\mathbf{D}_3$ are determined by the three reflections in $\mathbf{D}_3$. Suppose we let S be a reflectional symmetry in $\mathbf{D}_3$. Then the axis of symmetry L corresponding to S consists of those points z in the plane that are fixed by S, that is, those points z satisfying $Sz = z$.

Now suppose z is a point on the axis of symmetry determined by the reflection S. Since F is assumed to be $\mathbf{D}_3$ symmetric, we must have $F(Sz) = SF(z)$. But since z lies on the axis of symmetry determined by S, we have $Sz = z$. Consequently, $F(z) = F(Sz) = SF(z)$, and so $F(z)$ is fixed by S. Hence, $F(z)$ lies on the axis of symmetry determined by S.

Using the three axes of symmetry of $\mathbf{D}_3$, we may divide the plane into six wedge-shaped regions. We assign a color to each of these wedges. We refer to Figure 5.1, where we have shown a picture of the

Figure 5.4 Color coding of the eighth iterate.

computer screen with colored wedges displayed. Note that we have singled out a square region in the center of the screen with a black border.

We choose a point z inside each pixel lying in the square region. We then color that pixel with the color of the wedge in which $F(z)$ lies. There is an ambiguity concerning this construction which we mention here only briefly. The phenomenon of sensitive dependence on initial conditions leads to the possibility that two points inside one pixel will go under iteration by F to two different wedges. In our case, we have chosen the point to be the top left-hand corner of the pixel. Other choices will lead to essentially the same results. We show the results of this construction in Figure 5.2.

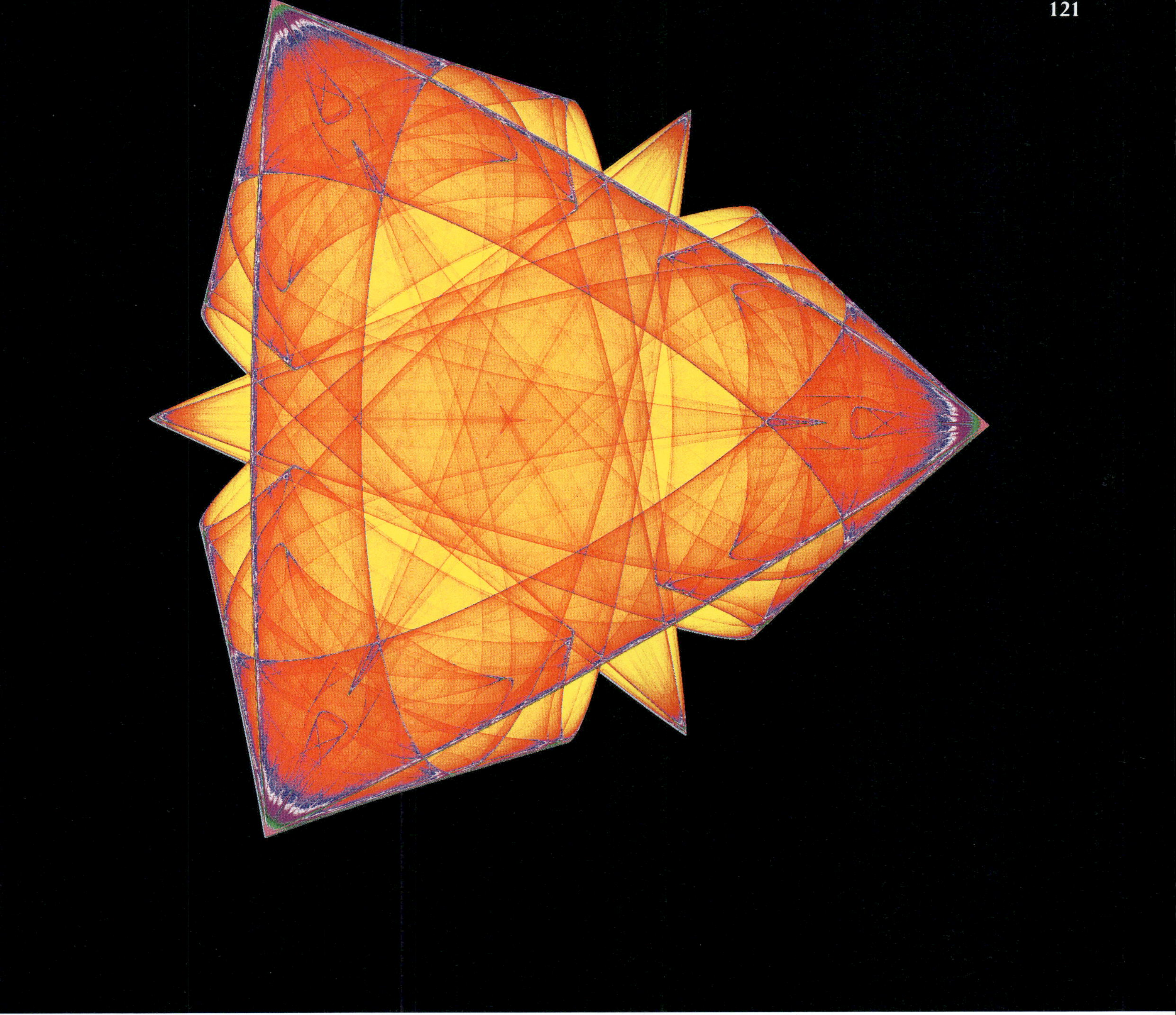

Figure 5.5 Golden Flintstone. Attractor for color coded mapping.

Observe that there is some evidence of mixing in Figure 5.2. For example, the coloring shows that points inside of any wedge get spread around by F to four different wedges. This kind of mixing is shown even more graphically in Figures 5.3 and 5.4. In Figure 5.3 we color code the fourth iterate $F(F(F(F(z))))$, and in Figure 5.4 the eighth iterate. You see immediately the effects of mixing in these pictures. Indeed, by the eighth iterate, we can choose six pixels that are quite close to each other and which go to all of the six possible wedges. This illustrates the phenomenon of sensitive dependence on initial conditions, and one can see how quickly it occurs.

In Figure 5.5, we show the symmetric attractor associated with the map F that we used to produce Figures 5.2–5.4.

Numerology

Our eventual goal is to present the formulas that we use to compute the symmetric icons. It turns out that these formulas become much simpler and more transparent once we have shown that points in the plane can be regarded in many ways like ordinary numbers. In particular, we want to show how it is possible to add and multiply points in the plane so that all the usual rules of arithmetic hold. However, before we do this, it is helpful to digress in order to sketch some fragments in the historical evolution of the idea of *number*.

As a guide, it is worth noting that two themes appear in this evolution. The first theme is that of utility or application. When all is said and done, numbers are used in computation, to solve problems. Over time, our concept of number has often been enlarged when it has become necessary to find a numerical solution to a problem. The other theme is more philosophical and mathematical in character. It is that of mathematical beauty or completeness. Often, the concept of number is enlarged so that mathematical problems that should reasonably have solutions do have solutions.

The remarkable fact is that both themes are inextricably intertwined. Often, new ideas of number have been developed to solve seemingly esoteric and abstract mathematical problems; at other times, new highly abstract ideas about numbers have evolved out of efforts by engineers and scientists to solve practical problems. Whatever the source, it has always turned out that abstract ideas of number are essential to our understanding of the real world.

By discussing certain aspects of the evolution of numbers —integers to fractions (or rational numbers) to irrational numbers to real numbers to complex numbers—we will see how the complex numbers, the points in the plane referred to above, may also be seen as a natural tool for computation. The books *The Emergence of Number* by John Crossley and *Number Theory, An Approach Through History* by André Weil present a more comprehensive discussion of the history of numbers.

The integers

The nineteenth century German mathematician Leopold Kronecker made the comment that the

natural numbers were made by God; the rest [of mathematics] is the work of man.

Noting Kronecker's remark, it seems appropriate to start our overview with the idea of a natural number. The natural numbers, or positive integers, comprise the *whole* numbers 1, 2, 3, ... As we have already indicated, the development of the idea of number is very much driven by practical considerations. For example, we need to extend our idea of whole number to include negative numbers just to do accounting using modern notions of debt and credit. Negative numbers (and zero) were certainly well known, understood, and used by the seventh century Indian mathematician Brahmagupta and his followers. On the other hand, negative numbers were treated with considerable distrust as late as the sixteenth century in Europe, probably due to their lack of geometric interpretation.

We call the set of all whole numbers—positive, negative and zero—the *integers*. However, integers by themselves are far too limited a class of number even for the analysis of problems just involving integers. We recall that a fraction (or *rational number*) is a quotient of integers. Thus, $\frac{1}{2}$, $-\frac{2}{3}$, and $\frac{16}{7}$ are fractions. Fractions appear the moment we look at averages. For example, the *average* size of a household in Australia in 1986 was about 2.9. This is so, even though adults and children come in whole numbers.

The rational numbers

The idea of number was important to Greek mathematics and philosophy. If we go back to the fifth century BC, the followers of Pythagoras—the Pythagoreans—were the dominant force in Greek mathematics. From our perspective, the Pythagoreans' idea of number is not easy to understand; roughly speaking we believe that they worked with natural numbers and fractions. Most of their ideas about number were closely related to geometry—to quantities like length, perimeter, and area. As a result there was no developed concept of negative numbers since such geometric quantities are always positive! On the other hand, fractions or ratios are closely tied to geometric constructions. For example, in Euclid's treatise on geometry, we find a construction, using ruler and compass, for the division of a line into n equal parts for any natural number n. It follows that, given a line of unit length, it is possible to construct a line of length m/n for any positive integers m and n.

To the considerable dismay and consternation of Greek mathematicians, it turned out that not all geometry could be expressed in terms of ratios of whole numbers. Referring to Figure 5.6, we recall the Pythagorean theorem that the square of the hypotenuse of a right triangle is the sum of the squares of the adjacent sides. (It is amusing to note that the question of who first proved the Pythagorean theorem is still a matter of some controversy. What does seem to be generally agreed is that it was not Pythagoras.)

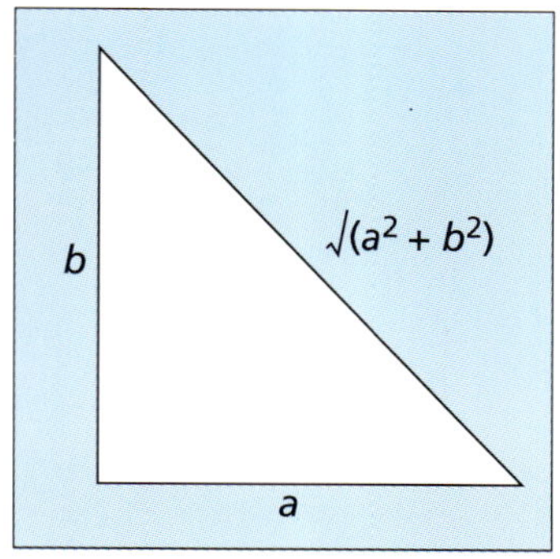

Figure 5.6 *The Pythagorean theorem.*

If the adjacent sides of a right triangle both have length 1, then it follows from the Pythagorean theorem that the length of the hypotenuse is $\sqrt{1^2+1^2} = \sqrt{2}$. However, as the Greeks were able to show, $\sqrt{2}$ cannot be written as a fraction, that is, $\sqrt{2}$ is an *irrational number*. This fact demonstrates that we cannot do geometry using only rational numbers.

The irrational numbers

The Pythagoreans regarded this discovery as rather shocking. Indeed, not only was their geometry based on the idea that all numbers were rational, but it is believed that their proof of the Pythagorean theorem depended crucially on this assumption. For some time the existence of irrational numbers was kept secret by the Pythagoreans. There is the story, circulated by the Pythagoreans, of the man who revealed the secret of irrational numbers and how he was drowned in a shipwreck.

In fact, the existence of irrational numbers *is* shocking. It is not by chance that we use the word 'irrational' to describe numbers like $\sqrt{2}$. If we attempt to write down the decimal expansion of $\sqrt{2}$, we find that

$$\sqrt{2} = 1.4142135623730950488016887 24 \ldots$$

and the sequence of integers following the decimal point never repeats itself. Consequently, we need *infinitely many* integers to specify precisely the square root of 2. In other words, we can never know $\sqrt{2}$ exactly. All we can say is that $\sqrt{2}$ is that positive number which squares to 2. Much later, it was shown in 1761 by the German mathematician and philosopher Johann Heinrich Lambert that the ratio π of the circumference to the diameter of a circle is also an irrational number. In one sense the existence of irrational numbers is not too serious. Indeed, a computer computes using rational numbers; it could never have a memory big enough to store an irrational number like $\sqrt{2}$. Nevertheless, in spite of this limitation, a computer often produces answers that are very close to the true, possibly irrational, result. From the philosophical and mathemat-ical point of view, however, we have to accept the existence of irrational numbers, and once we do this we find that *most* numbers are irrational. If we use decimal expansions, then we may write any number as an infinite decimal. For example,

$$\tfrac{1}{35} = 0.02857\,142857\,142857\,142857\ldots$$

The sequence 142857 repeats forever and this *periodicity* in the decimal expansion is characteristic of rational numbers. Indeed, rational numbers, and only rational numbers, have this periodicity in their decimal expansion.

Suppose now that we randomly choose each digit in the decimal expansion of a number. What sort of number would we expect to create? It seems most unlikely that there would be a pattern of digits that would repeat infinitely often; indeed it is most unlikely that a number chosen in this fashion would be rational. In this sense, most numbers are irrational.

The real numbers

The collection of all numbers, rational and irrational, is called the *real numbers*. There is some irony in the choice of the word *real*, as at each stage in the development of number new types of number, such as negative numbers or irrational numbers, were regarded with suspicion and given names such as *false* or *fictitious* numbers.

As we have just discussed, the existence of rational and irrational numbers were forced on the Greeks when they tried to do geometry, that is, to make sense out of concepts like length and area. There is another direction in which these types of numbers are also forced to exist. Try solving the linear equation $2x - 3 = 0$. The solution is easily seen to be $x = \frac{3}{2}$, which is not an integer. Thus, when you try to solve linear equations with integers as coefficients, you can easily get rational numbers—not integers—as solutions.

Next consider the slightly more complicated equation: the quadratic equation $x^2 = A$ when $A \geqslant 0$. This equation always has a solution, if we are allowed to use irrational numbers. For example, if $A = 2$, the solutions are $\pm\sqrt{2}$. Significantly, if $A > 0$ is irrational then the solutions of $x^2 = A$ will also be irrational. Thus, no new numbers are needed when we use the Pythagorean theorem to solve for the length of the hypotenuse of a right triangle given the lengths of the adjacent sides, since this solution involves taking the square root of a positive real number. Now consider solving the equation $x^2 = -A$ when A is positive. For example, consider solving the equation $x^2 = -1$. Such an equation has no real number as a solution, since the square of a real number is never negative.

In most formal treatments, a new number i is introduced to *solve* the equation $x^2 = -1$. That number is written formally as $i = \sqrt{-1}$. Even the notation seems to advertise this approach as strange—for the letter i stands for *imaginary*. We shall shortly show that there is a more natural and geometric way to think of i. Moreover, this alternative approach not only removes the mystery from i but also suggests and permits powerful new methods of computation. For the moment, however, note that our formal definition enables us to solve the equation $x^2 = -A$ when A is positive: the solutions are

$$\pm\sqrt{-A} = \pm\sqrt{-1}\,\sqrt{A} = \pm i\sqrt{A}.$$

More generally, we might ask when the quadratic equation $x^2 + 2bx + c = 0$ has a solution. Most readers will recall from high-school algebra that the solutions to this equation are given by the *quadratic formula* $-b \pm\sqrt{b^2 - c}$. Using the notion of i described above, we see that this formula works just fine. When $b^2 \geqslant c$, the square root produces a real number and the solutions are real. When $b^2 < c$, the square root produces a multiple of i. However, guided by potential applications, we might say, in this case, that equations where $b^2 < c$ have no conceivable application and dismiss them as aberrations. Mathematically, however, it seems quite unsatisfying that some quadratic equations have solutions while others do not. Historically, this problem did not worry mathematicians: solutions of quadratic equations were always thought about geometrically (not algebraically) and an equation $x^2 + 2bx + c = 0$ with $b^2 < c$ was simply regarded as an equation without solutions or geometric interpretation.

The complex numbers

Numbers involving i are called *complex numbers*; all complex numbers can be written in the form $x + iy$ where x and y are real numbers. As the quadratic formula shows, every quadratic equation has solutions which are either real numbers (when $b^2 \geqslant c$) or complex numbers (when $b^2 < c$). As we have seen, from the algebraic point of view, trying to solve linear equations leads to the need for rational numbers and trying to solve quadratic equations leads to the need for irrational and even complex numbers. There is a sinking feeling that if we try to solve cubic equations, then we will need even more exotic types of number. But that is not the case.

In the sixteenth century, mathematicians started to think seriously about solutions to cubic equations, specifically equations of the form

$$x^3 + ax = b.$$

The Italian physician and mathematician Girolamo Cardano was the first person to publish a formula for the solution of cubic equations. This formula is now named after Cardano though he learnt of it from a contemporary Niccolo Tartaglia and swore, under oath, not to disclose it. In fact another Italian, Scipione dal Ferro, is generally credited with being the first person to solve general cubic equations. Whatever the truth of the matter, Cardano presents a formula for the solutions of cubic equations in his famous book *The Great Art or the Rules of Algebra* and, for the first time, describes numbers including the square root of a negative number. However, he regarded these new numbers somewhat disdainfully:

> So progresses arithmetic subtlety the end of which, as is said, is as refined as it is useless.

In fact, complex numbers were far from useless. Rafael Bombelli, who was an engineer and contemporary of Cardano, worked out the basic rules of arithmetic for complex numbers. Bombelli was not interested in whether or not they existed; for him these numbers were useful precisely because they enabled him to solve practical engineering problems. Bombelli was not concerned about the philosophy behind these numbers. His view was pragmatic: they worked.

In spite of Bombelli's enthusiasm, mathematicians of the time tended to treat complex numbers with considerable caution, often combined with mysticism. Thus, Gottfried Wilheim Leibniz described these numbers as

> . . . a fine and wonderful refuge of the divine spirit, as if it were an amphibian of existence and non-existence.

Notwithstanding this initial scepticism, mathematicians increasingly studied complex numbers. By the nineteenth century, Karl Friedrich Gauss had christened these numbers *complex numbers* and, perhaps unfortunately, this rather intimidating name persists to this day.

As we shall soon see, complex numbers can be regarded as an extension of the real numbers. Most importantly, the usual rules of arithmetic hold for complex numbers.

Geometry of the complex numbers

The first point to note about the complex numbers is that they are all points in a plane—the *complex plane*. The second point to observe is that you can add and multiply complex numbers in a way that is consistent with the addition and multiplication of real numbers. The third point concerns the complex plane itself: there are two coordinate systems that we can use to describe complex numbers—*cartesian* and *polar*. Addition is most easily defined in cartesian coordinates, while multiplication is most easily described in polar coordinates.

We begin our discussion of the arithmetic of complex numbers with cartesian coordinates and addition. Recall that complex numbers can be written as $z = x + iy$, where x and y are real numbers. Thus we can view the point z as the point (x, y) in the (x, y)-plane. The real numbers x and y are called the *cartesian* coordinates of z (see Figure 5.7). The real number x is called the *real part* of z and is denoted by Re(z), while the real number y is called the *imaginary part* of z and is denoted by Im(z).Observe that if $y = 0$ then the complex number z equals the real number x. In this way, the real number line may be viewed as the horizontal x-axis in the complex plane. The real number zero is viewed as the origin of the complex plane and is denoted by O in Figure 5.7.

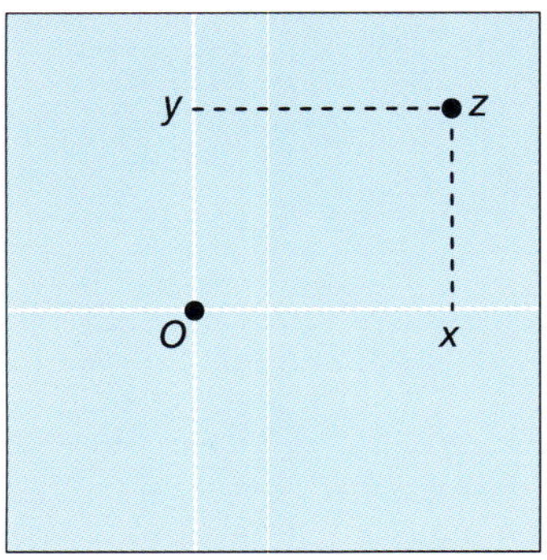

***Figure** 5.7 Cartesian coordinates of a complex number.*

Complex addition

To add the two complex numbers $z_1 = x_1 + iy_1$ and $z_2 = x_1 + iy_2$, we just add the real and imaginary parts separately, that is,

$$z_1 + z_2 = (x_1 + x_2) + i(y_1 + y_1).$$

Observe that if z_1 and z_2 are real ($y_1 = y_2 = 0$) then addition of complex numbers reduces to addition of real numbers.

Geometrically, the sum of two complex numbers is shown in Figure 5.8. Associate with z_1 and z_2 the parallelogram drawn in that figure. The sum $z_1 + z_2$ is the vertex of the parallelogram diametrically opposite the origin O.

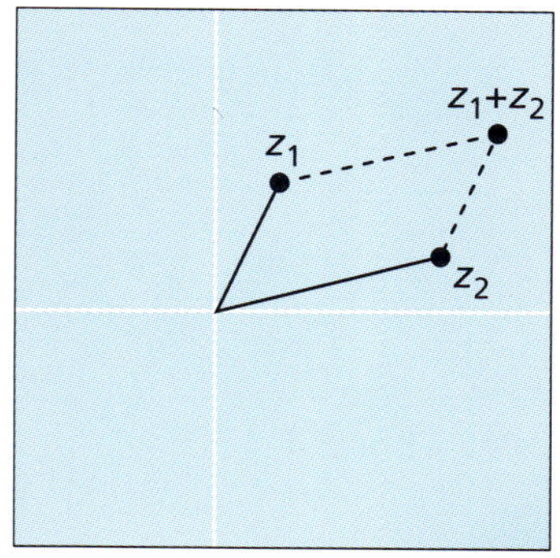

Figure 5.8 *Addition of complex numbers.*

Polar coordinates

Next we discuss polar coordinates. Let z be a complex number and let r be the distance of z from the origin. The number r is called the *magnitude* of z. The Pythagorean theorem implies that

$$r = \sqrt{x^2 + y^2}.$$

Now define θ to be the angle that the line through z and the origin makes with the horizontal x-axis, as shown in Figure 5.9. (If $z = 0$, we regard the angle θ as undefined.) The angle θ is called the *phase* of z, and the pair (r, θ) is called the *polar coordinates* of z. Observe that the real axis is also easily identified in polar coordinates: the positive real numbers have phase $\theta = 0°$, while the negative real numbers have phase $\theta = 180°$

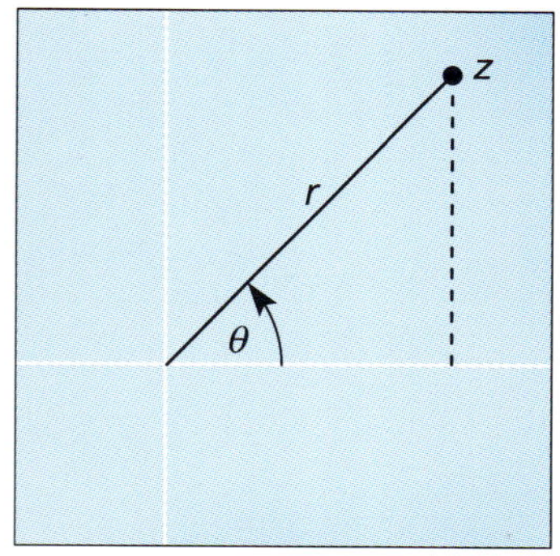

Figure 5.9 *Polar coordinates of a complex number.*

Complex multiplication

Using polar coordinates, it is rather easy to define complex multiplication. Suppose that z_1 and z_2 are complex numbers with magnitudes and phases (r_1, θ_1) and (r_2, θ_2), respectively. Then the product z_1z_2 is the complex number with magnitude r_1r_2 and phase $\theta_1 + \theta_2$ (see Figure 5.10). In polar coordinates, this rule for complex multiplication is simple: *multiply* the magnitudes and *add* the phases.

Suppose that z_1 and z_2 are the real numbers x_1 and x_2. Then the product z_1z_2 is also real and corresponds exactly to the real number x_1x_2. The phase of z_1z_2 is 0° if x_1 and x_2 have the same signs and is 180° if they have opposite signs.

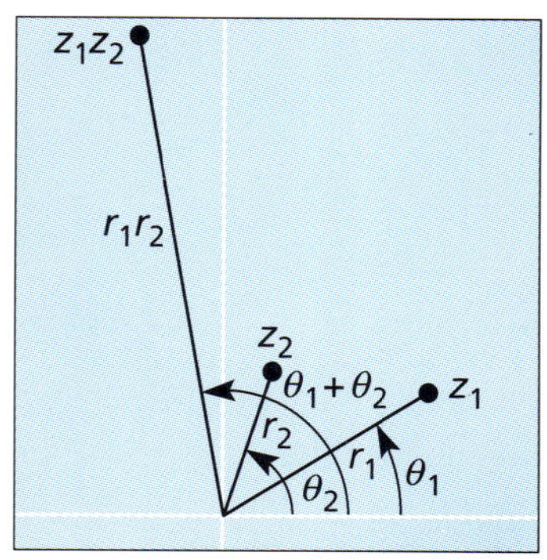

Figure 5.10 *Multiplication of complex numbers.*

As an example, consider the complex number z whose magnitude is 1 and whose phase is 90°. The number z^2 is a complex number whose magnitude is 1 and whose phase is 180°; that number is just the real number -1. Thus we have verified that z is a solution to the equation $z^2 = -1$; indeed, we have verified that the *imaginary* number i is just this number z. From this point of view, $i = \sqrt{-1}$ is a *fact*, not a *definition*.

One of the advantages of our definition of complex multiplication using polar coordinates is that it enables us to view rotation as multiplication by a complex number (of magnitude one.) As this interpretation of a rotation will be very helpful in our discussion of symmetry, it is worth taking some time to explain carefully what we mean. Suppose that ρ is a complex number of magnitude 1 and phase ψ. Then, multiplying the complex number z by ρ is the same as rotating z counterclockwise by the angle ψ. Indeed, the magnitude of ρz is the same as the magnitude of z and the phase of ρz is obtained by adding ψ to the phase of z. Thus the mapping

$$R(z) = \rho z$$

is rotation through the angle ψ.

In general, multiplication by a complex number ρ simultaneously rotates by the phase of ρ and dilates by the magnitude of ρ. Using this geometric fact, it is fairly easy to show that multiplication and addition of complex numbers satisfy all of the rules of ordinary multiplication and addition of real numbers.

We can now write down a formula for complex multiplication in cartesian coordinates. Suppose $z_1 = x_1 + iy_1$ and $z_2 = x_2 + iy_2$. Then

$$\begin{aligned} z_1 z_2 &= (x_1 + iy_1)(x_2 + iy_2) \\ &= x_1x_2 + ix_1y_2 + i_1x_2 + i^2y_1y_2. \end{aligned}$$

Using the fact that $i^2 = -1$, we see that

$$z_1 z_2 = (x_1 x_2 - y_1 y_2) + i(x_1 y_2 + x_2 y_1).$$

This somewhat forbidding looking formula has, as we now know, a natural interpretation as rotation and dilation. In cartesian coordinates, however, it is a simple formula with which to do computations.

Complex conjugation

There is one final operation on complex numbers that has proved to be most useful: *complex conjugation*. The complex conjugate of $z = x + iy$ is just $\overline{z} = x - iy$. In polar coordinates, $\overline{z}$ has the same magnitude as z but the opposite phase. More precisely, if $z = (r, \theta)$ in polar coordinates, then $\overline{z} = (r, -\theta)$ (see Figure 5.11). (Note that until now we have used positive numbers for angles, and these numbers refer to measuring angles in the counterclockwise direction. When we use negative numbers for angles, we are referring to clockwise measurement of angles.) It also follows that $z\overline{z}$ equals r^2, the square of the magnitude of z.

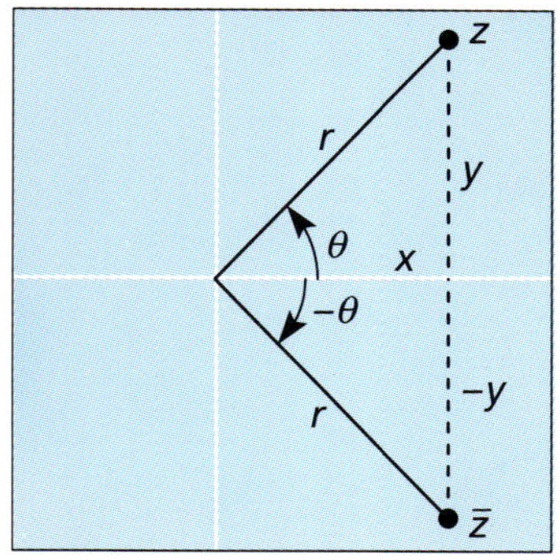

Figure 5.11 *The conjugate of a complex number.*

Finally note that geometrically the mapping

$$\kappa(z) = \overline{z}$$

is just reflection across the x-axis.

In a few pages, we have tried to summarize some of historical background to our contemporary ideas about number. These ideas quite literally took the greatest scientific minds thousands of years to develop. To conclude, we would like to emphasize two points. First, even for quite specific down-to-earth problems, we find that we are forced to extend and abstract our concept of number far beyond anything dreamt of by the mathematicians of antiquity. The situation was well summarized by Alfred Whitehead in 1925:

> The paradox is now firmly established that the utmost abstractions are the true weapons with which to control our thought of concrete fact.

Second, even though we tend to think of mathematics as understood and complete, at critical points in the development of mathematics and science mathematical ideas have been used that were not, in any sense, understood. The criterion was, and still is, pragmatic: if it works, use it! This was so with Bombelli's work on complex numbers and is true today with the physicist's use of Feynman integrals in quantum theory. It is this dichotomy between utility and understanding that has provided much of the driving force behind the development of science and mathematics.

Dihedral symmetry

Using complex multiplication, it is now an easy task to write down the symmetries of the dihedral group $\mathbf{D}_n$. Recall that the group $\mathbf{D}_n$ is generated by two elements: a reflection κ and a rotation R through $360°/n$. We may as well take the reflection to be reflection across the real x-axis, that is, in complex notation,

$$\kappa(z) = \bar{z}.$$

Suppose we let ρ be the complex number of magnitude 1 and phase $360°/n$. Then, using complex multiplication,

$$R(z) = \rho z.$$

The dihedral logistic mappings

Recall that one of our purposes in this chapter was to actually write down the formulas that we have used to produce the various symmetric icons. Now that we have introduced complex numbers and some of their properties, the presentation of these formulas is a relatively straightforward task. We begin by extending the odd logistic map to the plane so that it has the full symmetry of the circle. To do this, we merely define

$$G(z) = \lambda z(1 - z\bar{z}).$$

Write $z = x + iy$ and let (x, y) denote the cartesian coordinates of z. We now write the complex logistic map G in terms of x and y as

$$G(x, y) = \lambda\,[1 - (x^2 + y^2)]\,(x, y).$$

It is easy to see that if we set $y = 0$ we get the odd logistic map on the x-axis:

$$G(x, 0) = (\lambda\,(1 - x^2)x,\, 0) = (\,g_o(x),\, 0);$$

while if we set $x = 0$ we find the odd logistic map appearing on the y-axis:

$$G(0, y) = (0,\, \lambda(1 - y^2)y) = (0,\, g_o(y)).$$

Indeed, the odd logistic map actually appears on every line through the origin!

The remarks in the previous paragraph show that the extension of the odd logistic map to the plane is, in itself, not very interesting. Suppose we start iterating by G from some nonzero initial point z_0. We would find that all the iterates of this map would remain on the line in the plane going through the origin and z_0, and on that line the dynamics would be those of the odd logistic map described in the last chapter. In short, from a dynamics point of view there would be no reason to study this extended map.

What we want to do next is to modify this mapping so that, instead of having circular symmetry in the plane, it has only dihedral symmetry. To do this, we have to return to the idea of symmetry for a mapping introduced in a previous section.

Our previous discussion of the dihedral groups $\mathbf{D}_n$ recalls that the group $\mathbf{D}_n$ is generated by two symmetries: the flip κ and a counterclockwise rotation R of the plane through $360^\circ/n$.

We now observe that mappings F of the plane that have $\mathbf{D}_n$ symmetry are those that satisfy:

$$F(\kappa(z)) = \kappa F(z) \quad \text{and} \quad F(R(z)) = RF(z).$$

The explicit determination of mappings of the plane that are $\mathbf{D}_n$ symmetric is more difficult than the determination of mappings of the line which are odd. For the present, what we shall do is claim that the term of lowest degree which is $\mathbf{D}_n$ symmetric, but not circularly symmetric, is $\bar{z}^{n-1}$. It is easy to check that $F(\kappa z) = \kappa F(z)$, that is, $F(\bar{z}) = \overline{F(z)}$. In polar coordinates $F(z) = (r^{n-1}, -(n-1)\theta)$. It follows that

$$F(Rz) = \left(r^{n-1}, -(n-1)(\theta + 360^\circ/n)\right),$$

while

$$RF(z) = \left(r^{n-1}, -(n-1)\theta + 360^\circ/n\right).$$

The magnitudes of these two numbers are equal—but what about the phases. The difference between the phases is

$$\begin{aligned} & -(n-1)(\theta + 360^\circ/n) - [-(n-1)\theta + 360^\circ/n] \\ = & -(n-1)360^\circ/n - 360^\circ/n \\ = & -360^\circ. \end{aligned}$$

But any two angles that differ by 360° are the same. Hence the phases of $F(Rz)$ and $RF(z)$ are identical.

Adding the term $\bar{z}^{n-1}$ to G, we obtain the equation

$$g(z) = \lambda(1 - z\bar{z})z + \gamma \bar{z}^{n-1},$$

where λ and γ are real numbers. Maps of this form are the simplest mappings that can create symmetric icons in the plane through symmetry creation. We may think of these mappings as a natural extension of the (odd) logistic map to the class of $\mathbf{D}_n$ symmetric mappings.

For our exploration of symmetric icons, obtained through symmetry-increasing bifurcations, we have used a number of variations on this basic formula. In particular, most of the symmetric icons were obtained using the $\mathbf{D}_n$ symmetric mapping given by

$$F(z) = [\lambda + \alpha|z|^2 + \beta\,\mathrm{Re}(z^n)]z + \gamma\bar{z}^{n-1},$$

where λ, α, β, and γ are real numbers. The new term that we have added is $\beta\,\mathrm{Re}(z^n)\,z$; when $\beta = 0$ and $\alpha = -\lambda$ the mapping F is just the mapping g noted above. We have added this term because it is the $\mathbf{D}_n$ symmetric term of next highest degree to $\bar{z}^{n-1}$. In Appendix C, we give more details on how we actually find polynomial mappings with $\mathbf{D}_n$ symmetry.

Dihedral logistic maps in cartesian coordinates

When we perform the iteration process using the arithmetic rule F, we actually think of z as a point in the plane whose cartesian coordinates are x and y. We also write the complex number $F(z)$ in its real (u) and imaginary (v) parts, that is, $F = u + iv$. In this way, our formula is an arithmetic rule that shows us how to move the point (x, y) to the point (u, v). As noted previously, the brevity of the previous formula stems from the use of complex multiplication. To illustrate this point we shall write out the formula for F in cartesian coordinates when n equals 3 and 4. We begin by noting that

$$\bar{z}^2 = (x^2 - y^2) - i(2xy) \quad \text{and} \quad \mathrm{Re}(z^3) = x^3 - 3xy^2.$$

It follows that when F has triangular symmetry we can write $F(z)$ explicitly in cartesian coordinates as

$$F(x, y) = \big(I_3 x - \gamma(x^2 - y^2),\ I_3 y - \gamma(2xy)\big),$$

where

$$I_3 = \lambda + \alpha(x^2 + y^2) + \beta(x^3 - 3xy^2).$$

Similarly, when F has square symmetry, the formula for F in cartesian coordinates is

$$F(x, y) = (I_4x + \gamma(x3\text{-}3xy^2),\ I_4y - \gamma(3x^2y - y^3)),$$

where

$$I_4 = \lambda + \alpha(x^2 + y^2) + \beta(x^4 - 6x^2y^2 + y^4).$$

As n increases, writing out $F(z)$ in cartesian coordinates becomes ever more complex and we resist the temptation to give the formula for n greater than 4. Observe how much simpler it is to write out these formulas using complex numbers.

The use of complex notation is also helpful when we compute $F(z)$, since we actually compute $\bar{z}^{n-1}$ and $\mathrm{Re}(z^n)$ inductively using complex multiplication. This point is made more precise in Appendix B, where we present a program written in Basic for computing these icons.

More general $\mathbf{D}_n$ symmetric maps

For some of the symmetric icons, we have used a more complicated symmetric mapping that is not a polynomial. Specifically, we add a non-polynomial symmetric term to our standard $\mathbf{D}_n$ symmetric map F to obtain

$$F(z) = \{\lambda + \alpha z\bar{z} + \beta \mathrm{Re}(z^n) + \delta\, \mathrm{Re}([z/|z|]^{np})|z|\}z + \gamma z^{n-1},$$

where λ, α, β, γ, and δ are real numbers and $p \geqslant 0$ is an integer.

Often, the addition of this non-polynomial term has the effect of changing the symmetric icon near the origin, and we illustrate this phenomenon in Figure 5.12.

Figure 5.12 *The effect of adding a non-polynomial term:* (*a*) (opposite above) *with and* (*b*) (opposite below) *without non-polynomial term.*

$\mathbf{Z}_n$ symmetric maps

Until now, we have limited our discussion to mappings with full $\mathbf{D}_n$ symmetry. For both aesthetic and mathematical reasons, it is

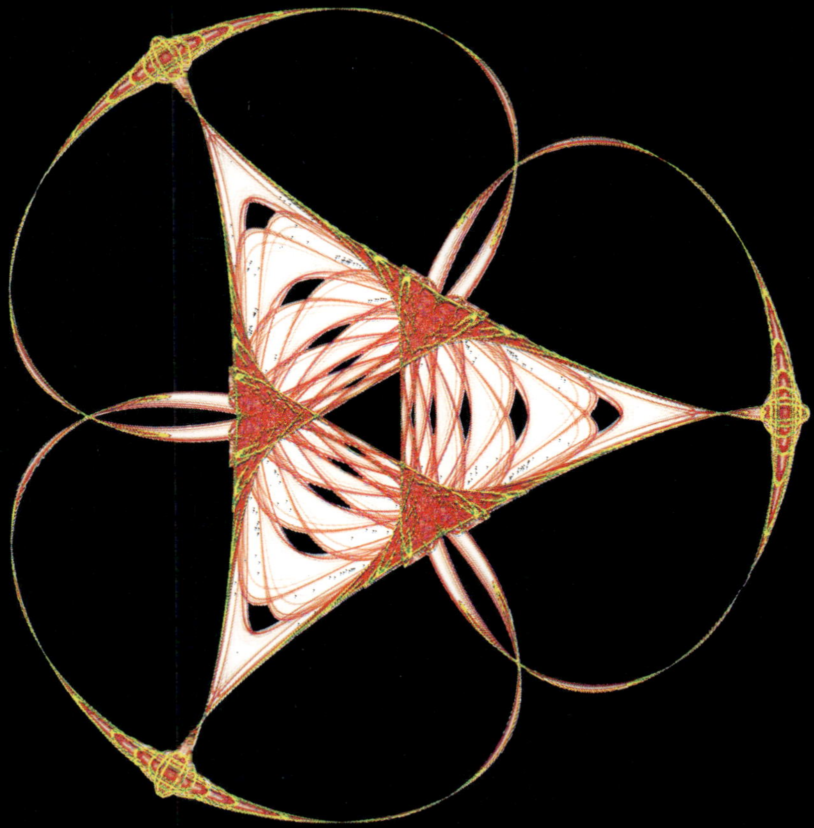

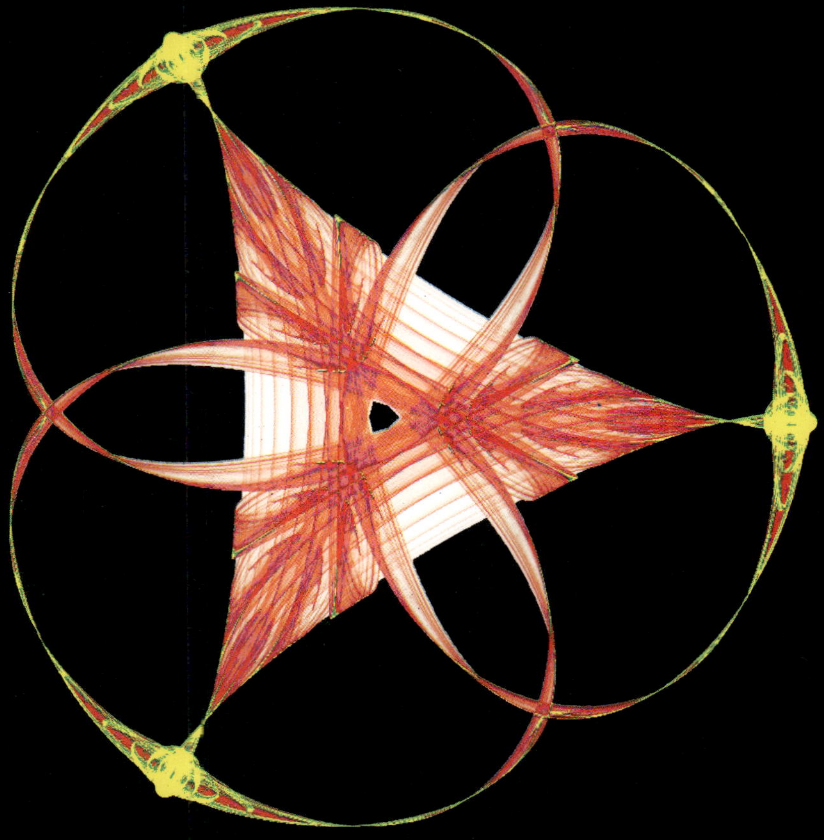

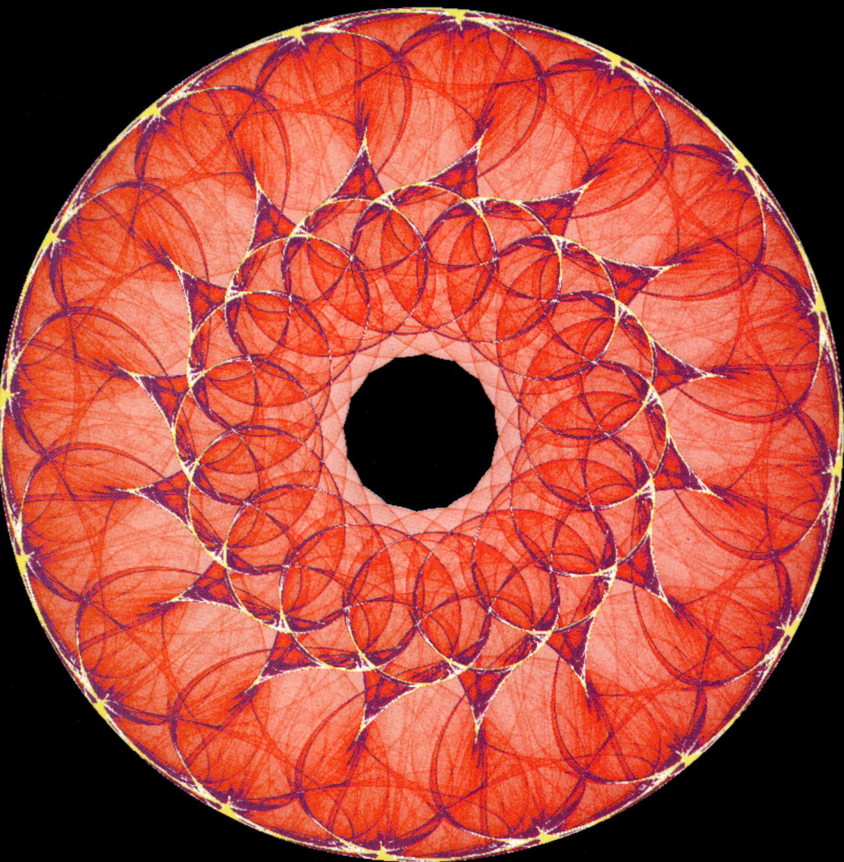

worth considering the effect of adding a term that breaks the reflectional symmetries in $\mathbf{D}_n$ while preserving the rotational symmetries in the cyclic group $\mathbf{Z}_n$. Such a term is ωiz. Observe that it is easy to show that $G(z) = iz$ does not have κ symmetry. For $G(\kappa z) = G(\bar{z}) = i\bar{z}$ while $\kappa G(z) = \overline{iz} = -i\bar{z}$. On the other hand, G does have rotational symmetry since

$$G(Rz) = G(\rho z) = i\rho z = \rho G(z) = RG(z).$$

The $\mathbf{Z}_n$ symmetric icon map that we explore is

$$F_{\mathbf{Z}}(z) = (\lambda + \alpha z\bar{z} + \beta\,\mathrm{Re}(z^n) + \omega i)z + \gamma\bar{z}^{n-1},$$

where λ, α, β, γ, and ω are real numbers. As the previous discussion shows, the mapping $F_{\mathbf{Z}}$ has the rotational symmetries generated by rotation through the angle $360°/n$ but does not have any reflectional symmetries, as long as ω is nonzero. It follows that the attractors obtained by using this map should not have a reflectional symmetry.

In Figure 5.13, we show the effect of the term ωiz on the attractor. The picture on the left shows the fully symmetric icon obtained when $\omega = 0$, while the picture on the right shows the attractor for $\omega = -0.15$. It is interesting to note that by adding a small amount of asymmetry to our mapping (ω is small), we were able to destroy the reflectional symmetries in the original attractor.

Figure 5.13 *Breaking the symmetry in an attractor from (a)* (opposite above) $\mathbf{D}_{16}$ *to (b)* (opposite below) $\mathbf{Z}_{16}$ *symmetry.*

Symmetry creation in the plane

We now discuss some of the consequences for the dynamics of $F(z)$ that follow from the existence of $\mathbf{D}_n$ symmetry. Suppose that S is any symmetry in the group $\mathbf{D}_n$. Since F is assumed $\mathbf{D}_n$ symmetric, we have

$$F(Sz) = SF(z).$$

As a consequence of this condition on F, we obtain a

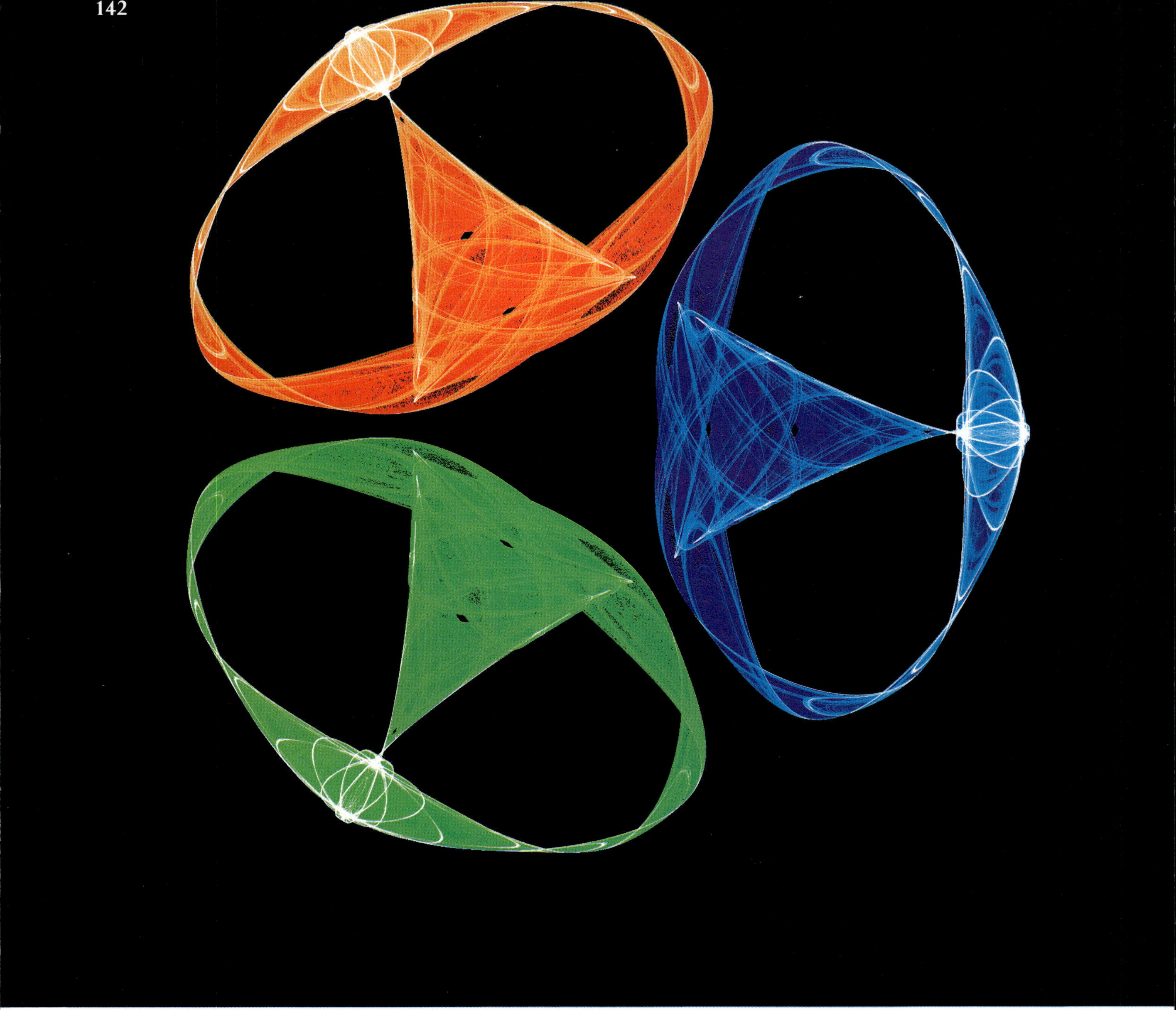

Figure 5.14 *Three conjugate attractors of a triangularly symmetric mapping.*

relation between the iterates computed at two symmetrically related points z and Sz :

$$z, \ F(z), \ F(F(z)), \ F(F(F(z))), \ldots$$

$$Sz, \ F(Sz), \ F(F(Sz)), \ F(F(F(Sz))), \ldots$$

It follows from symmetry that each point in the second sequence can be found by applying the symmetry S to the corresponding

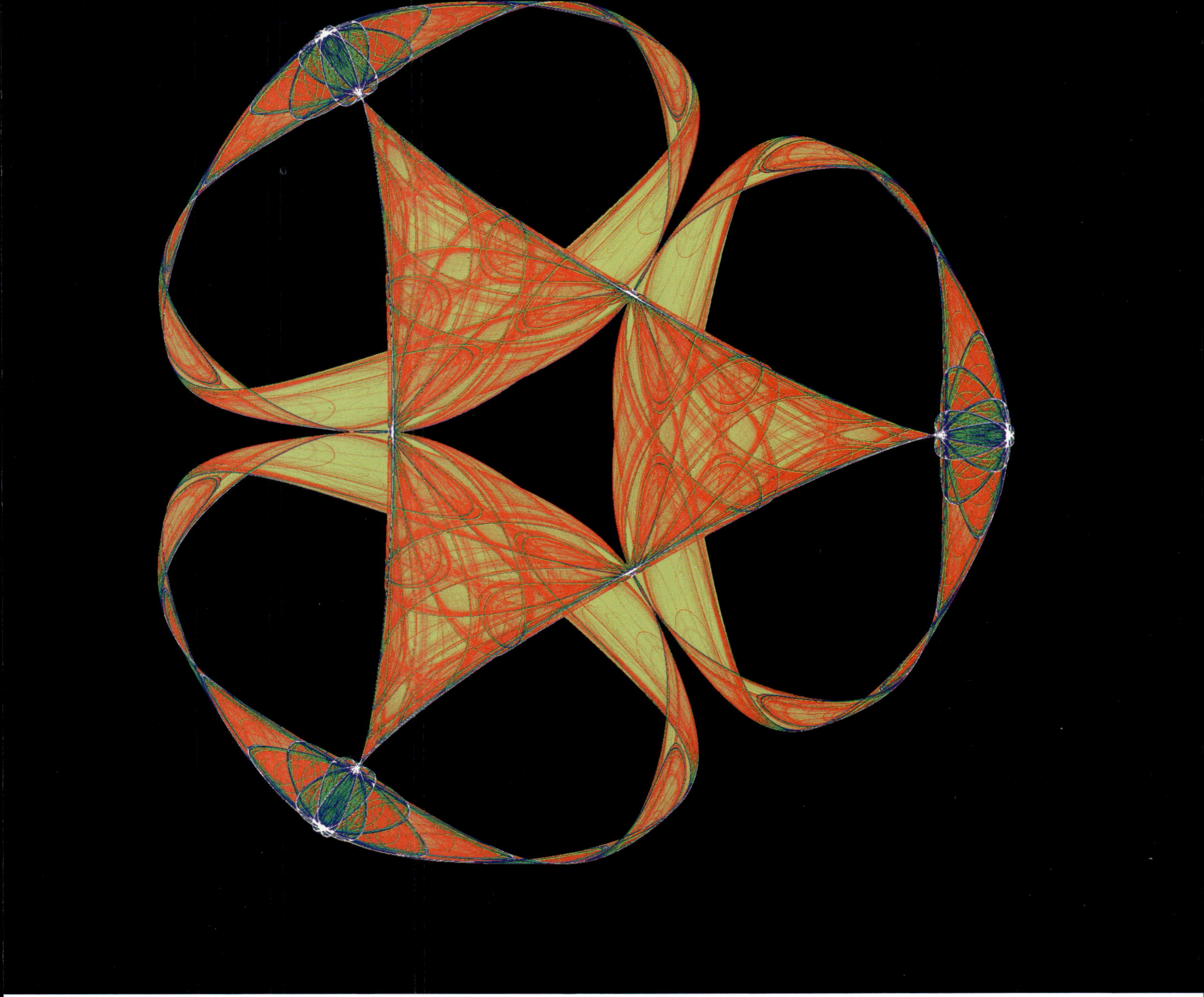

point in the first sequence. In Figure 5.14, we show the results of iterating a map with triangular symmetry starting with three symmetrically placed initial conditions, each indicated by its own color.

Figure 5.15 *Trinity in Red. Attractor with full triangular symmetry.*

Next, we discuss the consequence of symmetry for the pictures we obtain by using a symmetric mapping F. Suppose, we begin with an initial point z, iterate until the transients die out, and then plot a picture, which we label by A. As noted previously, if we began our iteration process with the point Sz, we would form the picture obtained by moving A using the symmetry S; we denote this picture by SA. There are two possibilities: either the pictures A and SA are indistinguishable, or they are not.

In the first case, A has the symmetry S. Indeed, in this volume we have already seen many pictures with nontrivial symmetries. In the second case, it can be proved that the pictures A and SA do not intersect and we call A and SA *conjugate* pictures. The three (color-coded) pictures shown in Figure 5.14 are all conjugate. We can use this discussion to understand how symmetry creation occurs in the plane.

Suppose that we change the constants in the mapping $F(z)$ by a small amount. The conjugate attractors such as those in Figure 5.14 may grow a small amount. Then, at a critical value of this constant, the conjugate attractors might actually *touch* along a line of symmetry. When this happens it can be shown that the new attractor will have the symmetry of the reflection corresponding to this line of symmetry. For example, each individual attractor in Figure 5.14 has a reflectional symmetry. When the attractors collide a second reflectional symmetry is added to the symmetries of the new attractor. Since the whole group $\mathbf{D}_3$ is generated by these two reflections, a new attractor with full $\mathbf{D}_3$ symmetry results (see Figure 5.15). This is our first example of symmetry creation in the plane.

Chapter six

QUILTS

WHEN we discussed planar symmetries in Chapter 2, we pointed out that there were two distinct types of planar symmetry group. First of all there are the finite planar symmetry groups. These groups contain only finitely many symmetries and arise by looking at the symmetries of geometric figures such as the square or n-sided regular polygon. A finite planar symmetry group is either a dihedral or a cyclic group. There is also the class of planar symmetry groups arising from the symmetries of repeating patterns in the plane. These groups, which we called wallpaper groups, contain infinitely many symmetries. As we have seen, pictures of symmetric chaos based on these two types of symmetry group lead to pictures that are quite different: the icons and the quilts. In the preceding chapter, we described the formulas that allow us to produce symmetric icons; in this chapter, we shall describe the methods by which we produce quilt patterns.

The quilts are made by a method similar to the one used to produce icons, but care must be exercised in creating patterns that will tile the plane in an aesthetically pleasing manner. The formulas that are needed to produce the quilts are more complicated than those used to create the icons and utilize trigonometric functions rather than polynomials.

Repeating patterns

Recently, while on a flight across half of America, one of us (MG) decided to take a small rather unscientific poll. What kinds of design were on men's shirts? More precisely, what kinds of design were on the shirts worn by the men travelling on this plane? Admittedly the largest

Figure 6.1 (opposite) *Gyroscopes.*

Figure 6.2 (opposite) *Cats Eyes. Quilt on a square lattice.*

class consisted of the rather boring (from the point of view of design) class of solid colored shirts, white and blue being the two most popular colors. Following the solids were the stripes—some broad, some narrow—but all with multicolored bands that repeated. Finally there were the plaids, usually formed by a combination of horizontal and vertical bands. That was the lot, apart from the one flamingo that had recently flown in from a Mexican vacation. The common feature of all of these patterns is that they can be repeated infinitely often. Indeed, in the era of mass produced textiles, the only patterns that are commercially viable are those that can be repeated *ad infinitum*.

The simplest of these repeating patterns are those built from a design on a single square whose horizontal and vertical translates fill up the plane. Indeed, all of the shirts—except, of course, the flamingo—were made from this single idea. Even the floor of the plane was covered by a repeating pattern of the same type. The carpet was a lovely maroon with a square array of little flesh-colored dots.

Our point is simple: once you start looking for repeating patterns, you see them everywhere. There are the patterns of tile on bathroom floors and kitchen walls, the hexagonal honeycomb pattern that honey-bees like to build, and the square pattern of backyard chain link fences. Thus, it seems natural, after finding so many different symmetric icons through symmetric chaos, to use that same technique to produce repeating patterns on planar lattices. To begin, we choose the square lattice.

The square lattice

Recall that the way we form patterns using chaotic dynamics is to choose a mapping, iterate while counting the number of times a point (pixel) is hit under iteration, and then color by number. To form a

repeating pattern, we need a way of assigning the same number to symmetrically placed points—in this case a lattice of symmetrically placed points. To achieve this symmetry, we cheat slightly. We iterate a mapping on a unit square to form the design on that square and then just tell the computer to fill out the plane (or as much of the plane as we desire) using translations of that design. We begin the discussion of the formulas that we use to make symmetric quilts by describing how it is possible to explicitly form mappings on a unit square. Denote the unit square in the plane by S. More precisely, a point $X = (x, y)$ is a point in S if the coordinates lie between 0 and 1, that is, if $0 \leqslant xy < 1$. Arithmetically there is a simple and natural way to translate points in the plane to points in S. Take a point (A, B) in the plane and write A and B as an integer part and a decimal; more precisely, write $(A, B) = (m, n) + (a, b)$ where m and n are integers and (a, b) is in the unit square S, that is, $0 \leqslant a,b < 1$. Geometrically the point (A, B) is translated to the point (a, b) in the unit square by this process.

Now we can write down mappings on S. Choose a planar mapping $f : \mathbf{R}^2 \rightarrow \mathbf{R}^2$. Suppose we apply the mapping f to a point (x, y) in the unit square. The result will be some point (A, B) in the plane. As described in the previous paragraph, we may translate this point to the point (a, b) in the square S. In this way we have created a mapping $\hat{f}$ on the unit square which takes the point (x, y) to the point (a, b). In symbols we write $\hat{f} : S \rightarrow S$.

Seams

As we indicated previously we will make quilt patterns from $\hat{f}$ by creating the design on S and then filling out the plane by translations of this design. There is a potential problem that we must address before beginning this process: *seams*. We don't want the pattern to have ugly seams at the boundaries of the square. To prevent seams from showing, we want our design on the right boundary of S to merge smoothly with the design on the left side of S; similarly, we want the design near the top boundary of S to merge smoothly with the design near the bottom side.

Mathematically, we can eliminate seams by demanding that the mapping f that we choose to start the process be consistent with the square lattice itself. More precisely, suppose we start with two points X and Y in the plane that translate to the same point in S. This happens when the components of X and Y differ by integers; that is, $Y = X + (m, n)$ where m and n are integers. For such points, we demand that $f(X)$ and $f(Y)$ also translate to the same point in S. That is, we demand that $\hat{f}(X) = \hat{f}(Y)$. In symbols, this means that there are integers M and N such that

$$f(X + (m, n)) = f(X) + (M, N).$$

If you look closely at the pictures of quilts, you will see that the choice of unit square is quite arbitrary. We might just as well have started with the square S' consisting of all points (x, y') where $0.11 \leq x' < 1.11$ and $1.43 \leq y' < 2.43$; the design on this square would appear to vary just as continuously as the design on the original square.

Internal symmetry

There is one difference, however between designs that we find on the two unit squares S and S'. The design that we have created on the square S has square symmetry, whereas the design on the square S' does not. As with the symmetric icons, we must choose the mappings f to respect the square symmetry of S in order to create designs with this internal square symmetry. We impose this square symmetry on the mapping $\hat{f}$ as follows.

Suppose that h is a symmetry of the square lattice. We demand that the map $\hat{f}$ commutes with this symmetry h. This is accomplished by requiring that, for all X,

$$f(hX) = hf(X) + (K, L),$$

where K and L are integers.

In Appendix D, we show how to derive mappings f that can produce designs on the unit square that are both seamless and square symmetric. The formula that produces these square quilt designs is somewhat complicated; we present that formula here for completeness:

$$\begin{aligned} f(x, y) = {} & m(x, y) + \mathbf{v} + \lambda(\sin 2\pi x, \sin 2\pi y) \\ & + \alpha(\sin 2\pi x \cos 2\pi y, \sin 2\pi y \cos 2\pi x) \\ & + \beta(\sin 4\pi x, \sin 4\pi y) \\ & + \gamma(\sin 6\pi x \cos 4\pi y, \sin 6\pi y \cos 4\pi x), \end{aligned}$$

where m is an integer, λ, α, β, and γ are real numbers, and either $\mathbf{v} = (0, 0)$ or $\mathbf{v} = (\frac{1}{2}, \frac{1}{2})$. Examples of quilts on a square lattice were given in Figures 6.1 and 6.2.

The hexagonal lattice

Of course, we could have required that our repeating pattern lie on any planar lattice; the square lattice was chosen first because it appears so frequently in tiles and quilts. For comparison we also produce repeating patterns based on the hexagonal lattice. Although the ideas used to create the mappings that produce patterns on the hexagonal lattice are identical to those that are used to create the mappings that

Figure 6.3 (opposite) *Flowers with ribbons.*

Figure 6.4 (page 154) *Sundial Mosaic. Quilt on Hexagonal Lattice.*

Figure 6.5 (page 155) *Fractured Symmetry.*

produce square quilt designs, the details are much more complicated. This formula is also derived in Appendix D. The mapping has four real parameters λ, α, and the vector $\boldsymbol{a} = (\beta, \gamma)$, and an integer constant m. The vectors $\boldsymbol{L} = (3, \frac{1}{\sqrt{3}})$, $\boldsymbol{M} = (2, 0)$ and $\boldsymbol{N} = (1, -\frac{1}{\sqrt{3}})$ are used in the definition of f. Below, R denotes rotation counterclockwise by 120° and F denotes the reflection of the plane that fixes the x-axis, that is, $F(x, y) = (x, -y)$. Finally, we use the dot product notation $\boldsymbol{N} \cdot \boldsymbol{X} = n_1x + n_2y$ where $\boldsymbol{N} = (n_1, n_2)$ and $\boldsymbol{X} = (x, y)$. The formula for f that we use is

$$\begin{aligned} f(\boldsymbol{X}) = m\boldsymbol{X} + \lambda[&\sin(2\pi\, \boldsymbol{N} \cdot \boldsymbol{X})\boldsymbol{N} + \sin(2\pi\, R\boldsymbol{N} \cdot \boldsymbol{X})R\boldsymbol{N} \\ &+ \sin(2\pi\, R^2\boldsymbol{N} \cdot \boldsymbol{X})R^2\boldsymbol{N}] + \alpha[\sin(2\pi\, \boldsymbol{M} \cdot \boldsymbol{X})\boldsymbol{N} \\ &+ \sin(2\pi\, R\boldsymbol{M} \cdot \boldsymbol{X})R\boldsymbol{M} + \sin(2\pi\, R^2\boldsymbol{M} \cdot \boldsymbol{X})R^2\boldsymbol{M}] \\ &+ \sin(2\pi\, \boldsymbol{L} \cdot \boldsymbol{X})\boldsymbol{a} + \sin(2\pi\, R\boldsymbol{L} \cdot \boldsymbol{X})R\boldsymbol{a} + \sin(2\pi\, R^2\boldsymbol{L} \cdot \boldsymbol{X})R^2\boldsymbol{a} \\ &+ \sin(2\pi\, F\boldsymbol{L} \cdot \boldsymbol{X})F\boldsymbol{a} + \sin(2\pi\, RF\boldsymbol{L} \cdot \boldsymbol{X})RF\boldsymbol{a} \\ &+ \sin(2\pi\, R^2F\boldsymbol{L} \cdot \boldsymbol{X})R^2F\boldsymbol{a}. \end{aligned}$$

An example of a quilt based on a hexagonal lattice is shown in Figure 6.4.

Less internal symmetry

Of course, as with the symmetric icons, the formulas can be modified so that the attractors have $\mathbf{Z}_4$ symmetry instead of $\mathbf{D}_4$ symmetry on the square lattice and $\mathbf{Z}_6$ symmetry rather than $\mathbf{D}_6$ symmetry on the hexagonal lattice. Examples of these are shown in Figures 6.3 and 6.5. Indeed, Figure 6.5 is found by adding a small symmetry-breaking term to the formula used to compute Figure 6.4.

Finally, we note that we could just as well have used a rectangular or a rhombic lattice on which to base our pictures. But from our perspective these lattices are less interesting, since they have less natural symmetry than the square and hexagonal lattices.

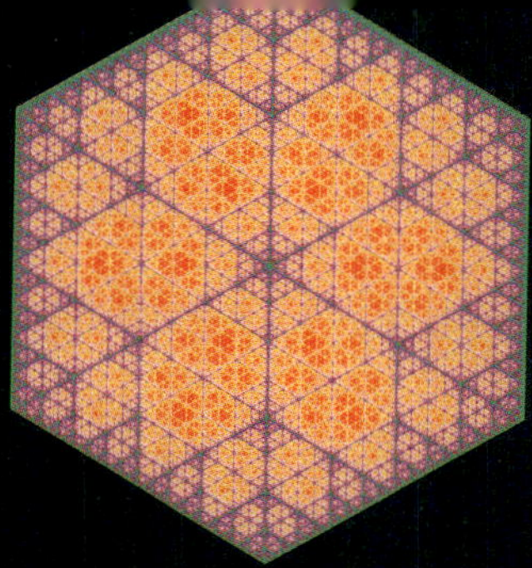

Chapter seven

SYMMETRIC FRACTALS

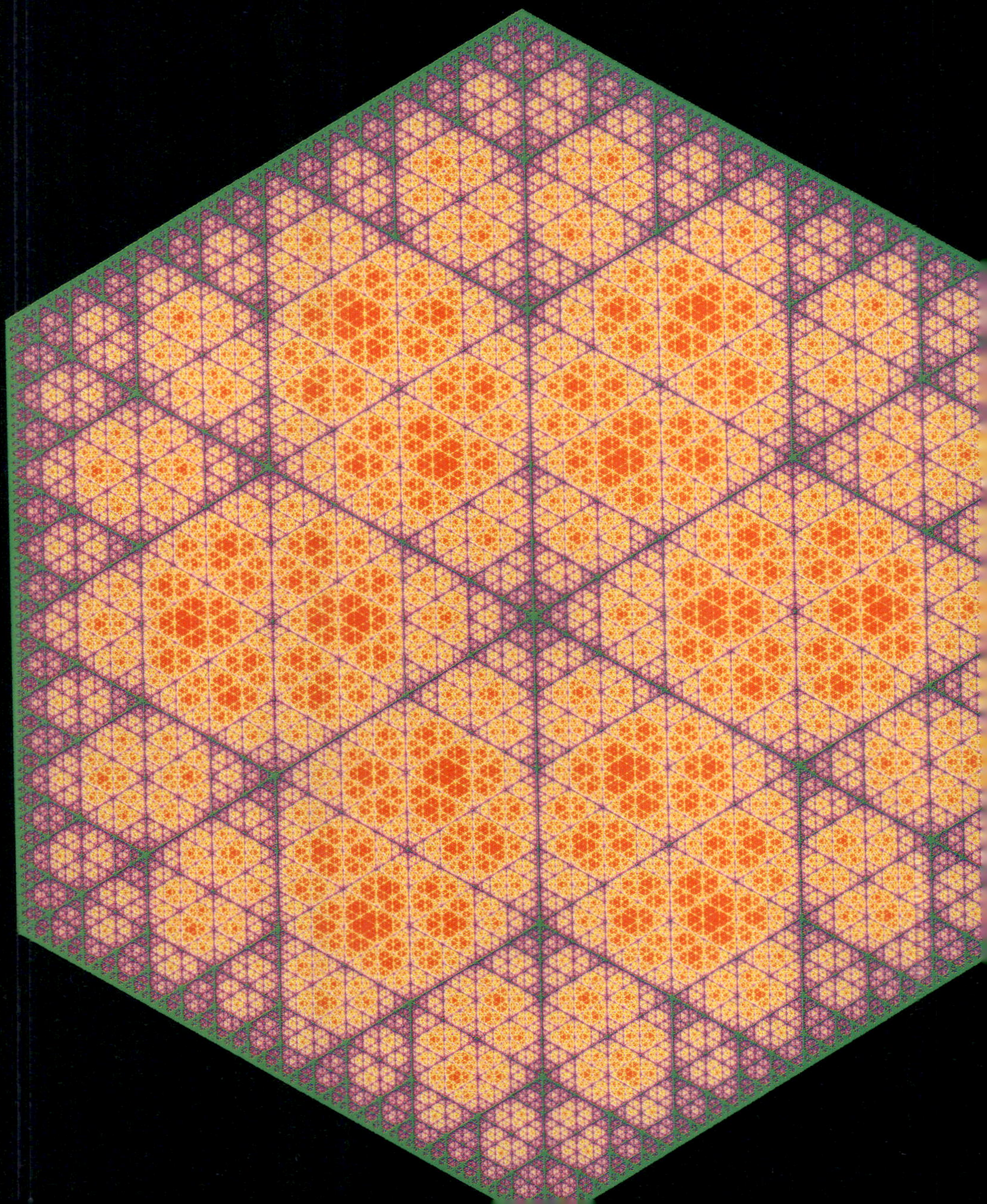

OUR method for constructing symmetric icons was based on choosing a fairly complicated symmetric polynomial mapping combined with the simple process of iteration. The result, as we showed, can be figures of great geometric intricacy. In this chapter, we adopt a rather different approach. Instead of working with one complicated mapping, we shall use a number of very simple mappings of the plane. Each of these mappings will have a single fixed point and when iterated will produce the corresponding point attractor. However, instead of using the process of iteration on a single mapping, we shall instead use a more subtle iterative process based on the *random* selection of mappings from our given set of simple mappings.

If we choose our set of simple mappings in a symmetric way, we obtain a symmetric figure in the plane. We call the resulting pictures *symmetric fractals*. In Figure 7.1 we show a fractal with 55-fold symmetry, which we call the *whipper-snipper*. As you can see, these symmetric fractals are strikingly different in feeling and texture from the symmetric icons, reflecting the fact that they are produced by methods quite different from those used to produce the icons and quilts.

Fractals

In recent years, there has been a veritable explosion of fractal-based computer art. There are the lovely books by Benoit Mandelbrot (*The Fractal Geometry of Nature*), Heinz-Otto Peitgen and Peter H. Richter (*The Beauty of Fractals*), and Michael Barnsley (*Fractals Everywhere*) showing some of this artwork. The methods for making fractal landscapes

Figure 7.1 (opposite) *Whipper-snipper.*

Figure 7.2 *A cartoon having a fractal theme by Sidney Harris*

and images have been used to create special effects scenes in a number of Hollywood movies, most notably perhaps in the *Star Wars* trilogy and in *Star Trek: The Wrath of Khan*. Fractals have also become commonplace enough to be the subject of cartoons in popular magazines and newspapers (see Figure 7.2).

Fractals have a curious mathematical property: they have essentially the same structure on all *scales*. This property was discovered and investigated in the early part of the twentieth century by a number of mathematicians. In 1915, the Polish mathematician Wrocław Sierpiński published the first pictures of what is now known as the Sierpiński triangle (Figure 7.3). If we examine the triangle, we see that it is composed of three smaller triangles, each of which is just a copy of the original Sierpiński triangle. In fact, if we magnify any one of the smaller triangles by a factor of two, we recover the original triangle. Sierpiński showed that this process may be repeated *ad infinitum*: each of the smaller triangles consists of three scaled down copies of itself. The Sierpiński triangle can be thought of as a mathematical instance of 'Plus ça change, plus c'est la même chose'.

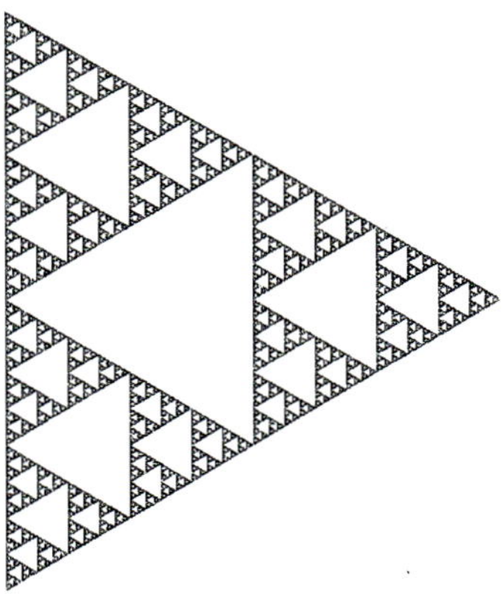

Figure 7.3 *The Sierpiński triangle.*

The Sierpiński triangle

Nowadays, we call figures which have this property of infinite reproducibility *fractals*. One might guess that the production of fractals like the Sierpiński triangle would require instructions of infinite complexity. However, one of the remarkable features of fractals is that they can be produced using very simple instructions or algorithms.

We illustrate this by describing the algorithm needed to construct a Sierpiński triangle. The novelty of these instructions is that they are based on choosing a sequence of *random numbers*.

We begin by choosing an equilateral triangle in the plane. Label the vertices A, B, and C and pick a point inside the triangle. Now choose randomly one of A, B, or C. If, say, B is chosen, move the point halfway to vertex B. Similarly, if A is chosen, move the point halfway to A, and the same for C. Now repeat the process to obtain a sequence of points that can be plotted on the computer monitor.

In this way we have defined a dynamical process, but a *non-deterministic* one. Nevertheless, if (after transients are ignored) we plot the points that are visited by this random process, we *always* (well actually with probability one) find the same figure appearing: the *Sierpiński triangle*.

That this procedure works and gives us the Sierpiński triangle seems quite remarkable. It is even more surprising when we contrast this method with that used to produce symmetric icons. In Chapter 1, we explained how a deterministic process can lead through the notion of chaos to something that looks random, and now we claim that a random process can lead to a figure of great regularity. You know the old adage 'You can't have it both ways'—well, in mathematics, sometimes you can.

Probability and random numbers

We digress now to discuss our use of the terms *randomly* and *with probability one* mentioned above.

We start with the concept of probability. In Australia,

there is a famous (and illegal) game called *two-up*. Two coins are taken and bets are placed on whether the coins, when tossed, will both fall the same side up. If we agree that one side of the coin is labelled the head, the other the tail, then bets are placed on whether the coins will fall both heads up or both tails up. If we assume that the coins are 'fair' (a moot point in illegal games), then the probability that the coins both fall tails up is $\frac{1}{4}$. This does not mean that if we toss the coins four times, then on one occasion the coins will fall tails up. The idea of probability is based on an average over time. To say that the probability that the coins will fall tails up is $\frac{1}{4}$ means that if we toss the coins a large number of times, say 100,000 times, then we can expect that approximately one-quarter of the coin tosses will result in both sides falling tail up. Note the use of the word *expect* it is possible, though *very unlikely,* that both the coins would fall tails up every single time. (Indeed, the odds against this happening are inconceivably large—about 10^{60000} to 1.) To say that the probability is $\frac{1}{4}$ means that, on average, we expect one in four coin tosses to result in a pair of tails.

Suppose that we want to choose a sequence of whole numbers between 1 and 10, say a sequence of 100,000 such numbers. How would we go about it? One possibility is to ask 100,000 people each to choose one whole number and record the results. Would we get a *random* choice by this method? The answer is almost surely no: numbers often have cultural associations there would be a bias towards choosing certain num-bers such as 3 and 7. So we repeat, how can we generate numbers between 1 and 10 in a 'random' way? One way of doing this, would be to imagine a ten-sided fair die, with each side marked with a whole number between 1 and 10. We then generate random numbers between 1 and 10 by successively tossing the die and taking the upmost number on the die. After many tosses, we would expect that each number should appear approximately the same number of times. Otherwise said, the probability of a given num-ber between 1 and 10 appearing uppermost should be $\frac{1}{10}$. In practice, it is difficult to construct a ten-sided fair die and instead we use a computer to generate random numbers between 1 and 10, or indeed between 1 and any whole number.

In the case of the Sierpiński triangle, we choose random numbers between 1 and 3. If we choose 1, we select the vertex A, 2 the vertex B, and 3 the vertex C. On average, in this way, we will choose the vertices A, B, and C approximately the same number of times. We can imagine, however, that it might happen that we choose the vertex B on each iterate. Then the resulting figure (after transients have been ignored) will be precisely the vertex B, not the Sierpiński triangle. When we say that the probability is one that the Sierpiński triangle will appear on the screen, we are implicitly claiming that the probability that vertex B will always be chosen is very small.

Indeed, we are saying more. Suppose that we call a sequence of vertices a *bad* sequence if it leads to a figure other than the Sierpiński triangle. Then we are asserting that, if the vertices are chosen at random, there is virtually no chance that a bad sequence will actually be chosen. Since there are many bad sequences—an infinite number of them—how is this possible?

Probability one

Suppose now that we change the question about choosing numbers between 1 and 10 to allow the choice of any number, integer or not. What will happen if we again ask a large number of people to choose a number between 1 and 10. Now it's less certain how 'most people' will reply, so to get a feel for the answer, we surveyed a few of our friends. What we found is that several of them continued to choose integers, but some more adventurous souls decided on fractions like $1\frac{7}{8}$ and $8\frac{2}{3}$, and others chose decimals like 3.14 and 5.28. It seemed as though people chose numbers that appeared to cover uniformly the numbers from 1 to 10. But were their choices *likely*? We begin by observing that the decimals that were chosen could equally well have been written as fractions. For example, $3.14 = 3\frac{7}{50}$ and $5.28 = 5\frac{7}{25}$. Next we recall from Chapter 5 that *all* fractions when written as a decimal have a special form: the decimal 'ends' in a pattern of digits that repeats itself *ad infinitum*. In the simplest of cases $7\frac{1}{2} = 7.500000$, that is, the decimal ends with the digit 0 being repeated infinitely often. Conversely, if a decimal has the property that it ends in a repeating pattern, then it can be written as a fraction. For example, the decimal

$$1.0588235294117647\ 0588235294117647\ 05882352\ldots$$

is equal to the fraction $1\frac{1}{17}$. Numbers, such as $\sqrt{2}$, that do not have this property are called irrational numbers.

From this point of view, it begins to seem rather unlikely that a number chosen randomly between 1 and 10 will be a fraction. If we think in terms of choosing the decimal expansion, then for the number to be a fraction we would have to choose the decimal expansion so that it ended with a pattern of repeating digits. Not only would such a choice be *unlikely*, it would almost never happen if we constructed the decimal expansion of a number by choosing each digit (an integer between 0 and 9) randomly. In this sense, we say that the probability that a number chosen at random between 1 and 10 will be a fraction is zero. It then follows that the probability that a number chosen at random between 1 and 10 will be irrational is one. In an analogous fashion, it can be shown that the probability that the iteration algorithm described above will lead to the Sierpiński triangle is one.

Sierpiński polygons

Our algorithm for constructing the Sierpiński triangle was based on fixing an equilateral triangle or, more precisely, fixing the vertices of that triangle. It is natural to modify the algorithm a little by replacing the triangle by a square or indeed by any one of the regular polygons. For example, suppose we had chosen a regular pentagon in the plane. After choosing an initial point inside the pentagon, we could have defined our algorithm by making random choices of vertices and, at each stage of the iteration, moving halfway towards the chosen vertex. In Figure 7.4, we show the figures that are obtained when we choose the regular pentagon and hexagon (we discuss the case of the square later).

We now look a little more carefully at the instructions that we use to construct the Sierpiński triangle. We choose the equilateral triangle in the plane with vertices $A = (1, 0)$, $B = (\frac{1}{2}, \frac{\sqrt{3}}{2})$, $C = (-\frac{1}{2}, -\frac{\sqrt{3}}{2})$, and note that the center of this triangle is the origin.

Pick a point $X = (x, y)$. Suppose that the vertex A is chosen. We move the point X halfway to A by averaging the coordinates of A and X. This leads to the map

$$\rho_A(x, y) = (\tfrac{1}{2}x + \tfrac{1}{2}, \tfrac{1}{2}y).$$

Similarly, since B has coordinates $(-\frac{1}{2}, \frac{\sqrt{3}}{2})$, we find that the map moving X halfway to B is given by

$$\rho_B(x, y) = (\tfrac{1}{2}x - \tfrac{1}{4}, \tfrac{1}{2}y + \tfrac{\sqrt{3}}{4}),$$

and the map moving X halfway to C by

$$\rho_C(x, y) = (\tfrac{1}{2}x - \tfrac{1}{4}, \tfrac{1}{2}y - \tfrac{\sqrt{3}}{4}).$$

It is worth remarking that the map ρ_A has the single fixed point A. If we iterate using only the map ρ_A, then successive iterates of X approach A. Indeed, A is an attractor for the iteration defined by ρ_A.

Figure 7.4 (*a*) (on following page 166) *The Sierpiński pentagon and* (*b*) (on following page 167) *the Sierpiński Hexagon.*

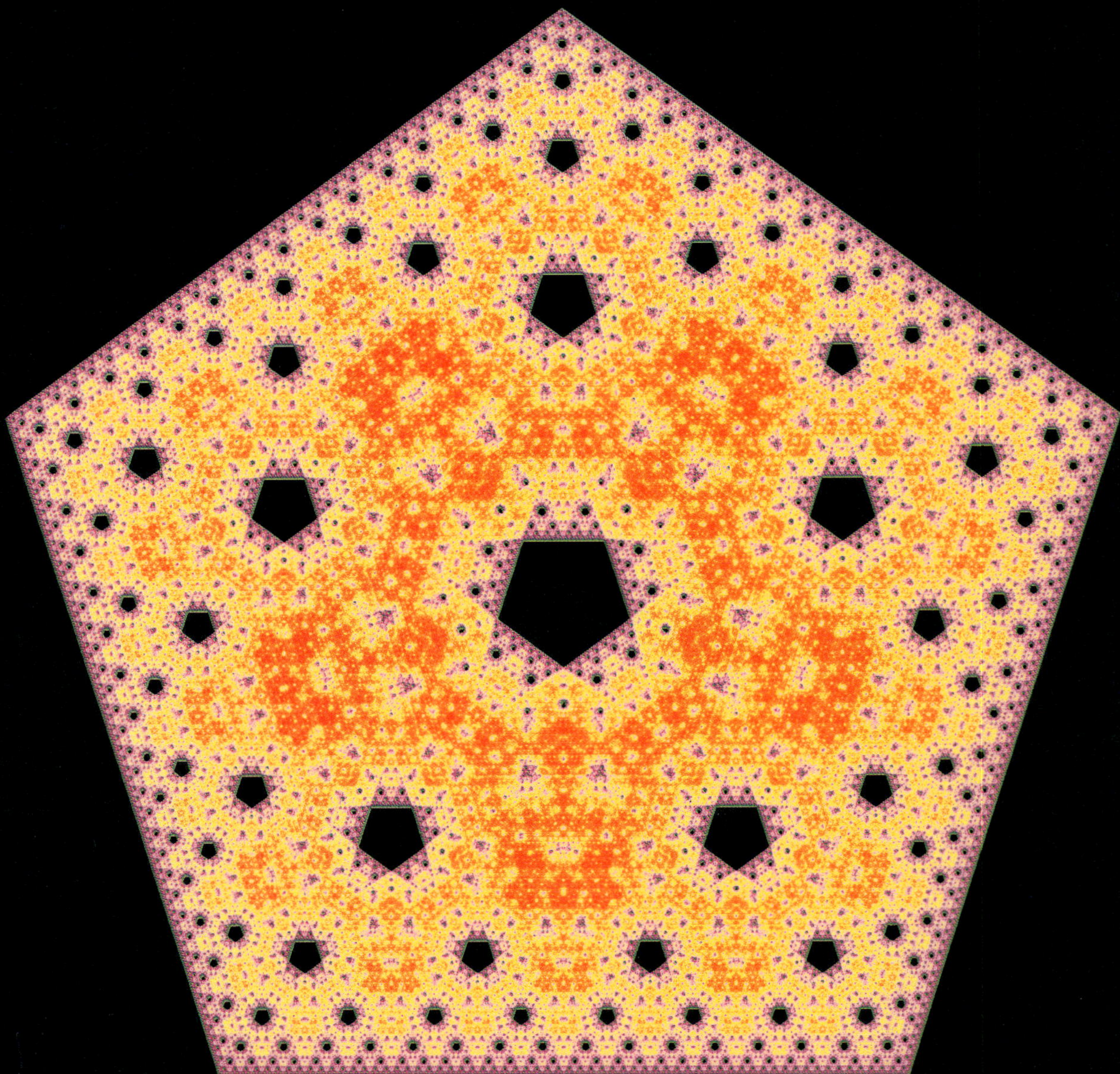

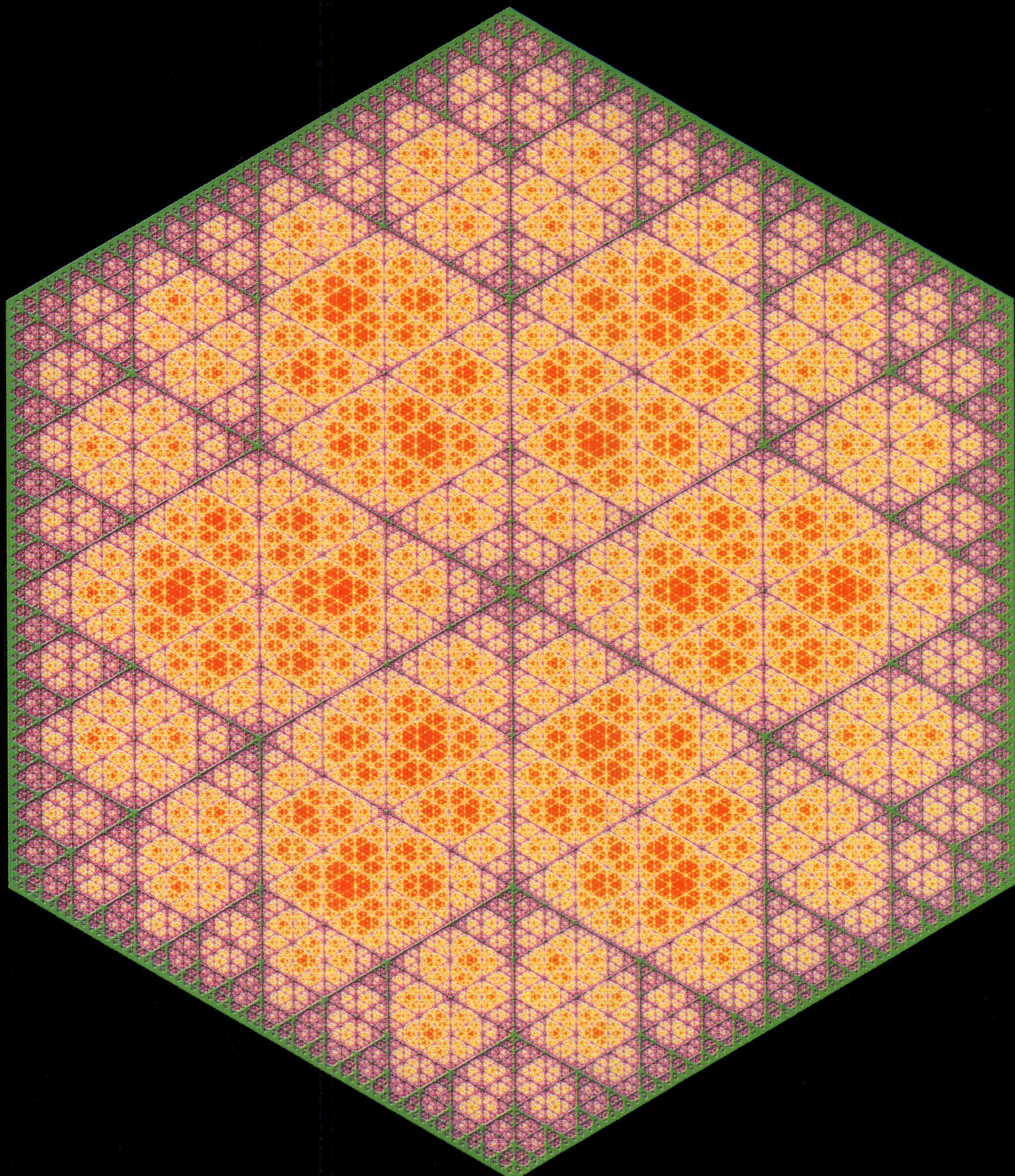

Similar remarks hold for the maps ρ_B and ρ_C. Taken individually, these maps have dynamical behavior akin to that of the 'halving' map. We conclude by emphasizing that the maps ρ_A, ρ_B, and ρ_C are very simple, especially by comparison with the symmetric polynomial maps we used to construct symmetric icons.

Affine linear maps

The mappings ρ_A, ρ_B, and ρ_C are examples of *affine linear* maps. In general, an affine linear map (of the plane) always has the form

$$\rho(x, y) = (ax + by + \alpha, cx + dy + \beta)$$

where a, b, c, d, α, and β are fixed real numbers.

In recent years, the method we have described for the construction of the Sierpiński triangle has been generalized to create a wide range of fractal images. We shall describe the main ideas that are used; the book by Michael Barnsley has more detail.

We start by choosing a finite collection of affine linear maps $\rho_1, \ldots, \rho_k$ and form a dynamical process similar to the one we described above in connection with the Sierpiński triangle. Specifically, choose an initial point X in the plane. Randomly select a whole number j between 1 and k and define the new point $\rho_k(X)$. This construction defines the iterative process. Provided the affine linear maps satisfy an additional condition, this process will, with probability one, converge to a fractal.

In fact, it should be clear that we must impose some conditions on the affine linear maps $\rho_l, \ldots ,\rho_k$. For example, suppose that the maps were all equal to the affine linear map $\rho(x, y) = (ax, ay)$, where a is a given real number. If a were greater than 1, then points would grow without bound under iteration. In fact, since we would always be choosing the same map, the process would be similar to the doubling map. On the other hand, if a were positive, but less than 1, points would converge to the origin and the process would be similar to the halving map.

Contractions

We restrict attention to maps of the plane which are *contractions*. To explain the idea of a contraction, suppose that ρ is a map of the plane and k is a positive number strictly less than 1. We say that ρ is a *contraction*, with *contraction rate* k, if given any pair of points X and Y in the plane, the distance between $\rho(X)$ and $\rho(Y)$ is less than or equal to k times the distance between X and Y. Basically, a contraction possesses the property that under iteration pairs of points are brought closer together—distances are scaled down by the factor k. The affine maps ρ_A, ρ_B, and ρ_C that make the Sierpiński triangle are all contractions since they halve distances to their corresponding vertices.

It turns out that any polynomial map of the plane that is a contraction *must* be an affine linear map. In particular, none of the maps we used to construct symmetric icons is a contraction. With a little more work, one can show that the affine linear map

$$\rho(x, y) = (ax + by + \alpha, cx + dy + \beta)$$

is a contraction if

$$a^2 + c^2 < 1,$$
$$b^2 + d^2 < 1,$$
$$a^2 + b^2 + c^2 + d^2 < 1 + (ad - cb)^2.$$

Iterated function systems and fractals

Barnsley calls a set $\rho_1, \ldots, \rho_k$ of affine contractions an

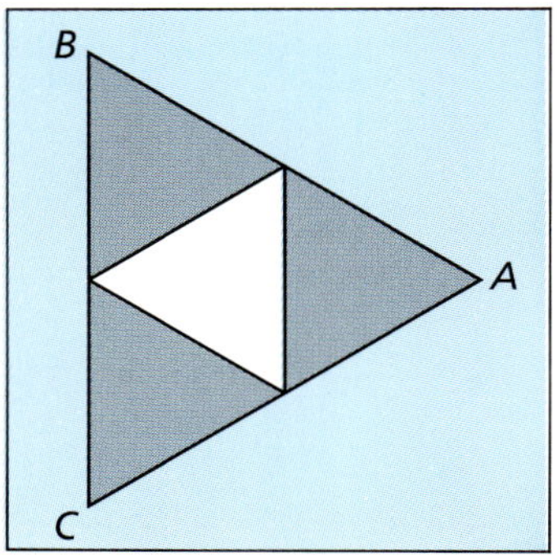

Figure 7.5 New triangles from old.

iterated function system. We can use the idea of an iterated function system to give a definition of a fractal. We shall say that a set L of points in the plane is a *fractal* if we can find an iterated function system $\rho_1, \ldots, \rho_k$ such that the points in L are precisely the points that lie in at least one of the sets $\rho_1(L), \ldots, \rho_k(L)$. Using some standard mathematical notation, we may express this more economically by writing

$$L = \rho_1(L) \cup \ldots \cup \rho_k(L).$$

We say L is the aggregate or *union* of the sets $\rho_1(L), \ldots, \rho_k(L)$. Since the map ρ_1 is a contraction, we can think of $\rho_1(L)$ as being a scaled down or *shrunken* copy of L. Similar remarks hold for the other sets $\rho_2(L), \ldots, \rho_k(L)$. Consequently, to say that L is a fractal means that it is the union of a (finite) number of scaled-down copies of itself.

It is possible to show that fractals always have the property of similar structure at all scales. This property *can* appear in a rather simple and uninteresting fashion, as we shall illustrate later. However, the type of structure we see in the Sierpiński triangle is characteristic of most fractals and we illustrate the idea of a fractal using the Sierpiński triangle.

As an example, let us look more carefully at what the contractions ρ_A, ρ_B, and ρ_C are actually doing and how they tend to scale down pieces of the triangle. To simplify our notation a little bit, we let S denote the equilateral triangle whose vertices are A, B, and C. The set $\rho_A(S)$ consists of all the points $\rho_A(X)$, with X lying in S. We similarly define $\rho_B(S)$ and $\rho_C(S)$.The contraction ρ_A scales S down by a factor of a half and it follows that $\rho_A(S)$ is an equilateral triangle with side of length half that of the original triangle A. Similar remarks hold for $\rho_B(S)$ and $\rho_C(S)$. In Figure 7.5, we show how the three triangles $\rho_A(S)$, $\rho_B(S)$, and $\rho_C(S)$ fit inside S. The equilateral triangle S pictured in Figure 7.5 is *not* a fractal since the set $\rho_A(S) \cup \rho_B(S) \cup \rho_C(S)$ is contained in S but is not equal to all of S: the set misses all points in the white central triangle in S.

With the goal of understanding how the Sierpiński triangle actually arises, we repeat this process but with S replaced by the set S_1 consisting of points which lie in one of the sets $\rho_A(S)$, $\rho_B(S)$, or $\rho_C(S)$. Symbolically, we have

$$S_1 = \rho_A(S) \cup \rho_B(S) \cup \rho_C(S).$$

Now $\rho_A(S_1)$ is just a scaled down version of S_1 as are $\rho_B(S_1)$ and $\rho_C(S_1)$. In Figure 7.6, we show the points in

$$S_2 = \rho_A(S_1) \cup \rho_B(S_1) \cup \rho_C(S_1)$$

We are still a long way from the Sierpiński triangle but there is now a definite resemblance.

Figure 7.6 *Towards the Sierpiński triangle.*

Again S_2 is not a fractal, because of the omission of yet smaller white triangles. We can continue this process indefinitely and in the limit reach a set S_∞. This set is the Sierpiński triangle and it has the property that the same kinds of white triangles are omitted on all scales. Not very surprisingly,

$$S_\infty = \rho_A(S_\infty) \cup \rho_B(S_\infty) \cup \rho_C(S_\infty),$$

and so S_∞ is a fractal.

Strict fractals and overlaid fractals

If the sets $\rho_1(L), \ldots, \rho_k(L)$ are *disjoint* (that is, they have no points in common), we say L is a *strict fractal*. The Sierpiński triangle is very close to being a strict fractal. For example, $\rho_A(S_\infty)$ and $\rho_B(S_\infty)$ have but one point in common (the mid-point of the edge AB of the original triangle). In general, in our definition of a fractal, we do not require the sets $\rho_1(L), \ldots, \rho_k(L)$ to be disjoint. In this case the fractal is *overlaid*. Most of the pictures of symmetric fractals that we show are of this type. In Figure 7.7, we show the effect of varying the rule for the construction of

Figure 7.7 Strict and overlaid fractals.

the Sierspiński triangle. In (a) we have defined ρ_A by moving a point 0.55 of the way to vertex A. The maps ρ_B and ρ_C are defined similarly. In (b) we move 0.50 of the distance to the chosen vertex and in (c) we move 0.45 towards the chosen vertex. As you can see, in (a) we obtain a strict fractal, in (b) we have the Sierpiński triangle (the transitional case), and in (c) we have an overlaid fractal.

Two basic theorems

In Barnsley's book, there are two beautiful theorems about iterated function systems. The first theorem states that every iterated function system $\rho_1, \ldots, \rho_k$ is associated with a unique fractal. That is, there is a unique bounded subset S such that

$$S = \rho_1(S) \cup \ldots \cup \rho_k(S).$$

The second theorem gives an effective way of constructing this fractal. If we start with a point in the plane and execute the dynamic random process that we have described above, then with probability one we form the fractal S.

At first glance, it may seem surprising that one can actually show the existence of a set S satisfying the conditions of the first theorem. Indeed, being able to make such an assertion illustrates one of the powers of mathematical thinking. Since the idea behind this proof is

Figure 7.8 (opposite) *Circular Saw*

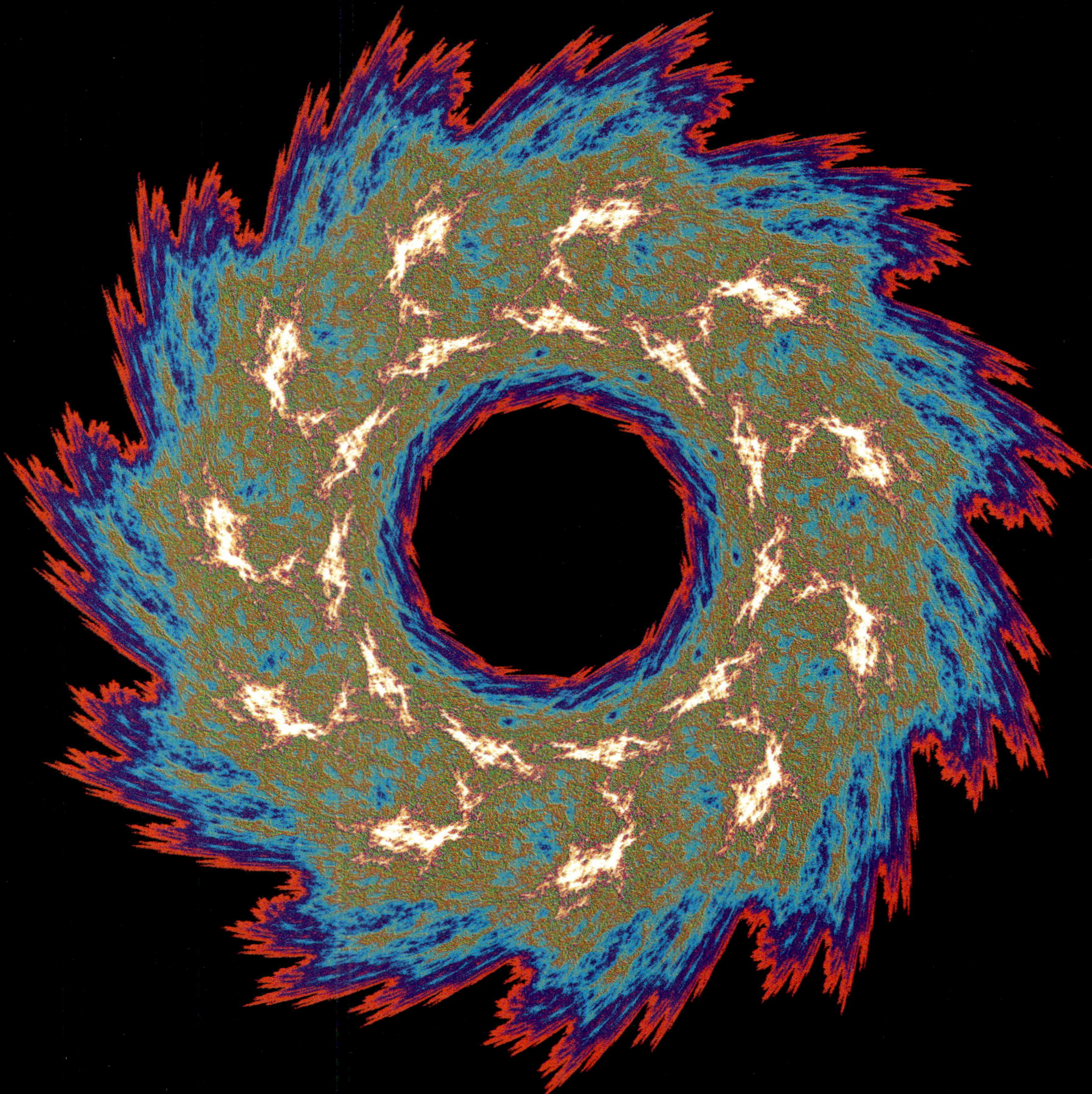

Figure 7.9 (opposite) *Catherine Wheel*

so very elegant, we shall attempt to give a short description of why the result is true.

One of the characteristic features of contractions is that they have a unique fixed point and that iterates of *any* initial point converge to that fixed point. The basic idea in the proof of the first theorem is to use the iterated function system to define a contraction on the set of all bounded subsets of the plane. This is clearly a sophisticated idea since it is not immediately clear how one would even define the distance between two planar sets. However, it is easy to define the mapping from subsets of the plane to subsets of the plane. If we are given an iterated function system $\rho_1, \ldots, \rho_k$ and a subset L of the plane, we define a new subset $\rho(L)$ of the plane by means of the formula

$$\rho(L) = \rho_1(L) \cup \ldots \cup \rho_k(L)$$

Indeed, for the Sierpiński triangle where we chose S to be the equilateral triangle, the set $\rho(L)$ is just the set S_1. It can then be shown that this transformation of subsets of the plane is a contraction mapping which has a unique fixed point. This fixed point is a set which we denote by S. Since S is a fixed point of ρ, we have

$$S = \rho_1(S) \cup \ldots \cup \rho_k(S),$$

and so S is a fractal satisfying the conditions of the theorem.

We see that the construction of the fractal S can be done in terms of an iteration. The difference with our construction of symmetric icons is that the transformation ρ is a contraction, and therefore simple, but the space on which it acts is complicated, and so we can end up with a geometric object of great complexity. Indeed, for the Sierpiński triangle, the iterates of ρ just form the sets S_k, and the limit set S_∞ is the Sierpiński triangle.

The method of proof of the first theorem suggests a way to construct the fractal associated with an iterated function system: take a

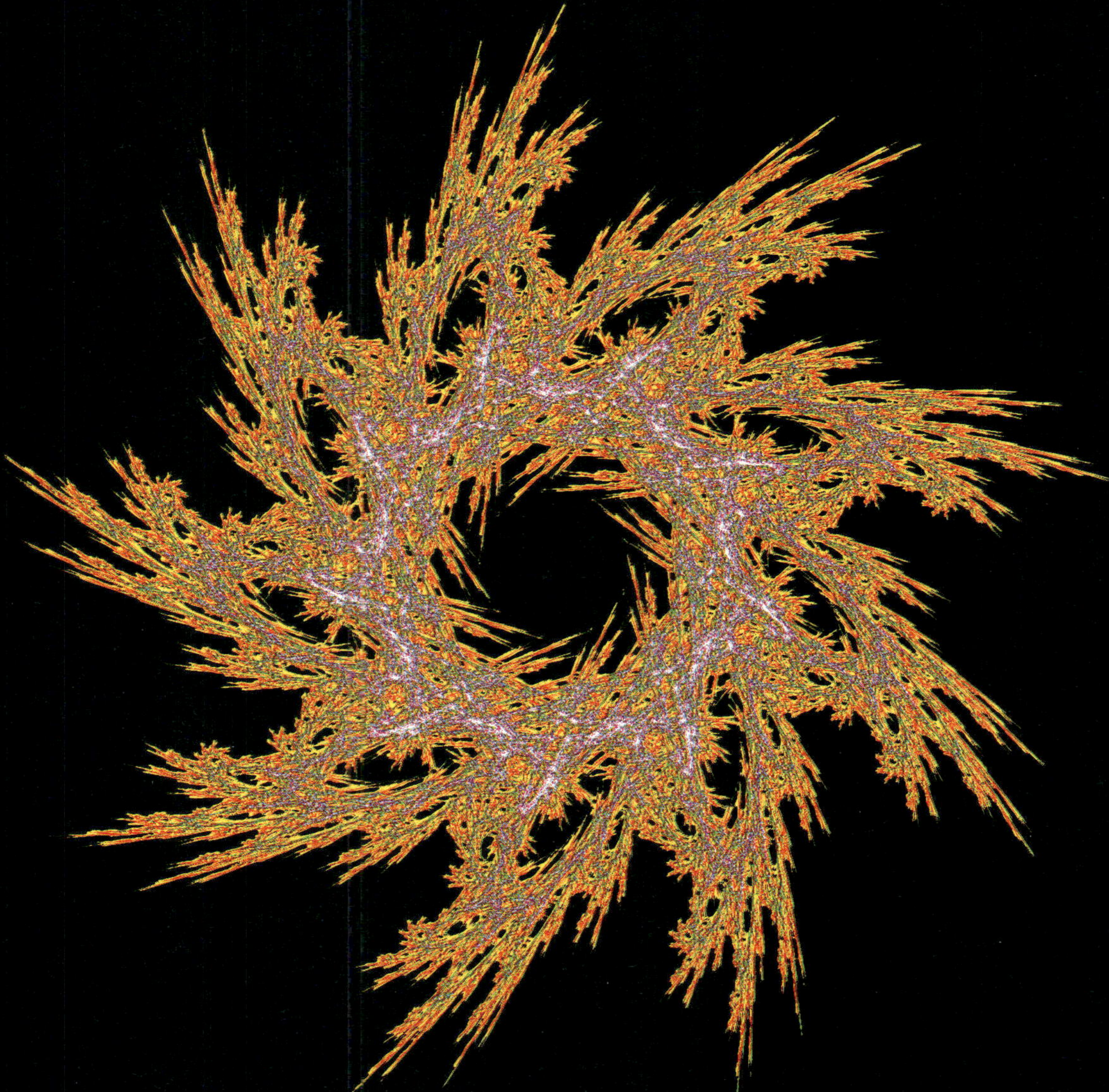

Figure 7.10 (opposite) *The Bee.*

subset S of the plane and iterate using the map ρ we described above. Theoretically this procedure is fine; practically it requires enormous computing resources. The gist of the second theorem is that we can construct the fractal merely by taking any point in the plane and carrying out the dynamical process described above. In somewhat fancier language, we can think of the fractal obtained by the first process as a *spatial* average and that obtained by the second process as a *time* average. The second theorem asserts that the spatial average and time average are equal. This result is an example of what is called an *ergodic* theorem. Although this subject is well beyond the scope of this text, suffice it to say that the idea of ergodicity is central to the modern theory of dynamical processes.

The Sierpiński square

As mentioned previously, there is one case where it is rather easy to describe the fractal that results when we use the first process. Suppose we take a square with vertices A, B, C, and D. We take as our iterated function system the affine contractions $\rho_A, \ldots, \rho_D$ defined by moving points halfway to the corresponding vertex. These are exactly the affine contractions we used in our discussion of the Sierpiński polygons. Suppose we denote the square by S and subdivide S into four squares, as shown in Figure 7.11. Observe that the set $\rho_A(S)$ is just the small square that contains the vertex A which is shown in black in this figure. Similarly, the sets $\rho_B(S)$, $\rho_C(S)$, and $\rho_D(S)$ are the small squares containing the vertices B, C, and D, respectively. Thus S satisfies

$$S = \rho_A(S) \cup \rho_B(S) \cup \rho_C(S) \cup \rho_D(S).$$

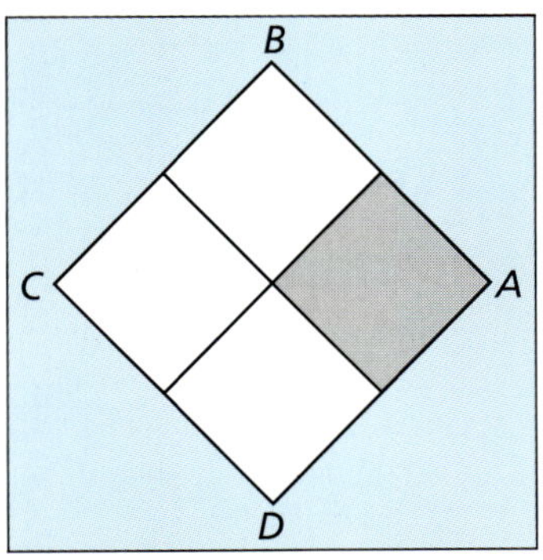

Figure 7.11 *Subdivision of the Sierpiński square.*

It follows that S is a fractal—a rather uninteresting fractal by comparison with the other Sierpiński polygons.

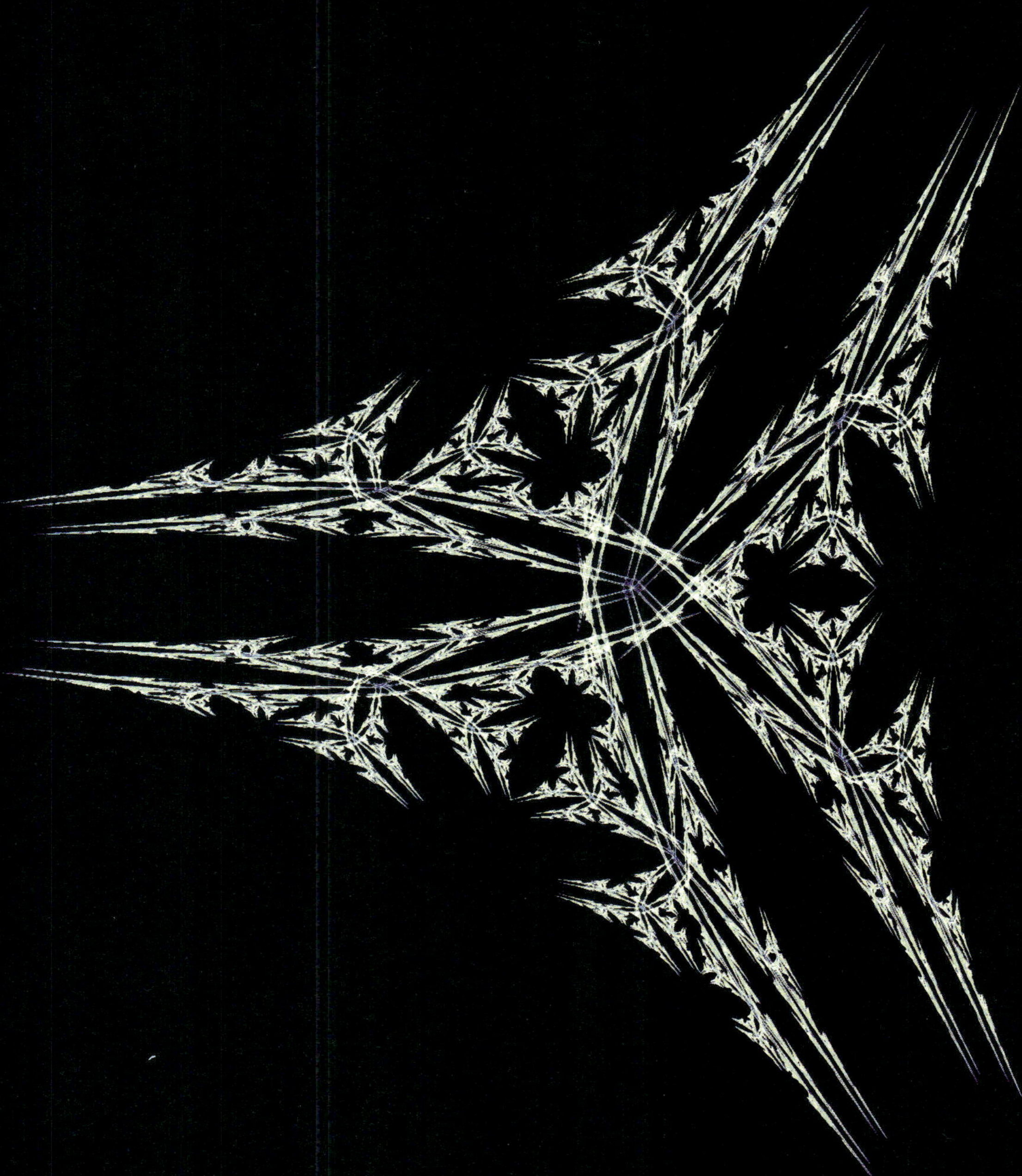

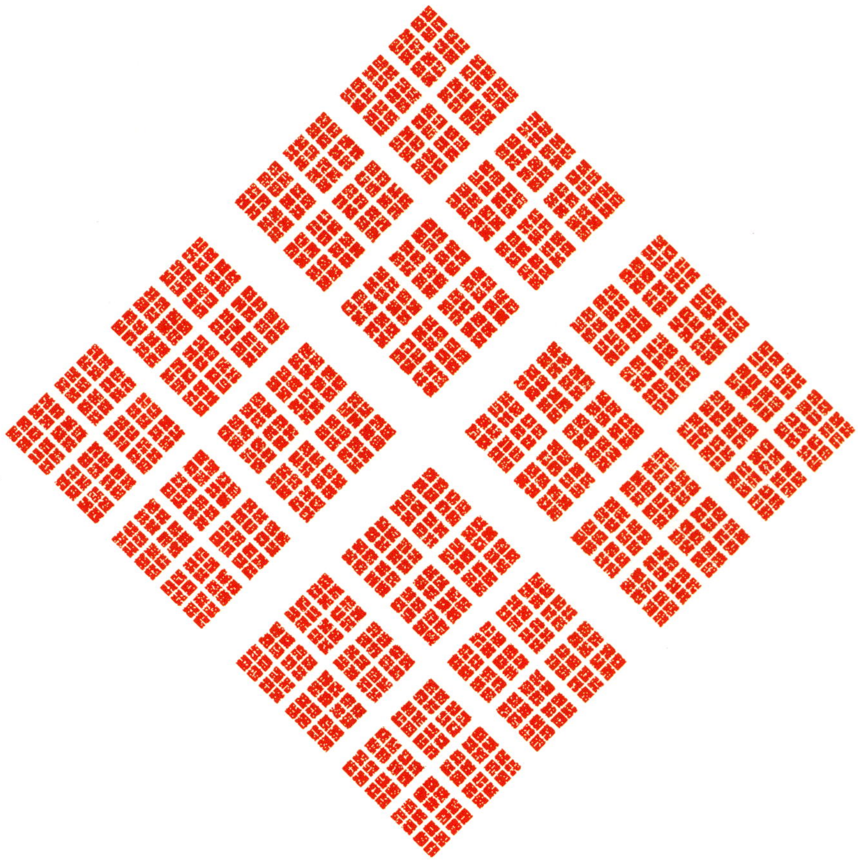

Figure 7.12 *Variation on the Sierpiński square.*

It is worth noting that for this construction to work the affine maps had to be chosen exactly. For example, in Figure 7.12, we show the fractal obtained if instead of moving halfway to the four vertices *A, B, C,* and *D*, we move only 0.46 of the way.

Symmetric fractals

We now show how we can use certain special kinds of iterated function system to form *symmetric fractals*. Fix an affine linear

Figure 7.13 (opposite) *The Doilly*

Figure 7.14 (opposite) *Astigmatism.*

contraction ρ. Suppose that G is a planar symmetry group equal to either $\mathbf{D}_n$ or $\mathbf{Z}_n$. For each element g in G, one can show that $g\rho$ is an affine contraction. Hence we may define an iterated function system consisting of the affine maps $g\rho$ with g an element of G. With this construction of an iterated function system, the resulting fractal has to be symmetric.

We illustrate this for the case when G is the cyclic group $\mathbf{Z}_3$. If we let r_{120} denote the rotation through 120°, then $\mathbf{Z}_3$ consists of the identity, r_{120}, and r_{240}. Consequently, if ρ is an affine contraction, the associated iterated function system consists of the affine contractions ρ, $r_{120}\,\rho$, and $r_{240}\,\rho$. It follows from the first theorem that there is an associated fractal S characterized by

$$S = \rho(S) \cup r_{120}\,\rho(S) \cup r_{240}\rho(S).$$

Observe that each element of the group $\mathbf{Z}_3$ just permutes the sets $\rho(S)$, $r_{120}\,\rho(S)$, and $r_{240}\rho(S)$. For example, r_{120} applied to these three sets is $r_{240}\rho(S)$, $r_{120}r_{120}\rho(S)$, and $r_{120}r_{240}\rho(S)$. Since $r_{120}r_{120} = r_{240}$ and $r_{120}r_{240}$ is the identity, the new sets are $r_{120}\rho(S)$, $r_{240}\rho(S)$, and $\rho(S)$. Thus $r_{120}(S) = S$ and the resulting fractal must have $\mathbf{Z}_3$ symmetry.

We use the second theorem to compute symmetric fractals. Suppose ρ is an affine contraction and G is a planar symmetry group as above. Suppose we have plotted the points $X_1, \ldots, X_n$. We plot the next point X_{n+1} by randomly choosing a group element g in G and taking X_{n+1} to be equal to the point $g\rho(X_n)$. By the first theorem, the set S *always* has G symmetry.

We now present some examples of symmetric fractals following the same scheme for coloring that we used for symmetric icons. In particular, the color of a pixel represents the probability that that pixel will be visited during the iteration (see the discussion of statistics in Chapter 1).

Figure 7.15 (opposite) *Paisley*

The Sierpiński triangle is symmetric

Although it is a fact that the Sierpiński triangle pictured in Figure 7.3 is $\mathbf{D}_3$ symmetric, the reason for this symmetry is slightly different from the reason that the symmetric fractals constructed in the last section are symmetric. As a consequence of untangling these differences, we shall see that two different iterated function systems can create the *same* fractal. Along the way, we shall also see that the first theorem—giving the existence of a unique fractal for every iterated function system—really has mathematical teeth.

We have already shown how the Sierpiński triangle may be constructed using the geometrically defined iterated function system ρ_A, ρ_B, and ρ_C. A consequence of the uniqueness part of the first theorem is that any bounded set of points Z in the plane satisfying

$$Z = \rho_A(Z) \cup \rho_B(Z) \cup \rho_C(Z) \qquad \text{(SIER)}$$

must be the Sierpiński triangle. Indeed, we will show that the Sierpiński triangle is actually symmetric by finding a symmetric set Z that satisfies (SIER).

We begin by finding a $\mathbf{Z}_3$ symmetric set. Suppose that r_{120} denotes rotation through $120°$. Then $\mathbf{Z}_3$ consists of the identity, r_{120}, and r_{240}. Using our recipe for constructing symmetric fractals, we consider the iterated function system ρ_A, $r_{120}\rho_A$, and $r_{240}\rho_A$. Associated with this iterated function system, there will be a unique $\mathbf{Z}_3$ symmetric fractal Z characterized by

$$Z = \rho_A(Z) \cup r_{120}\rho_A(Z) \cup r_{240}\rho_A(Z).$$

We shall show that Z satisfies (SIER) and so must be equal to the Sierpiński triangle.

Geometrically, the map $r_{120}\rho_A$ moves the point X halfway

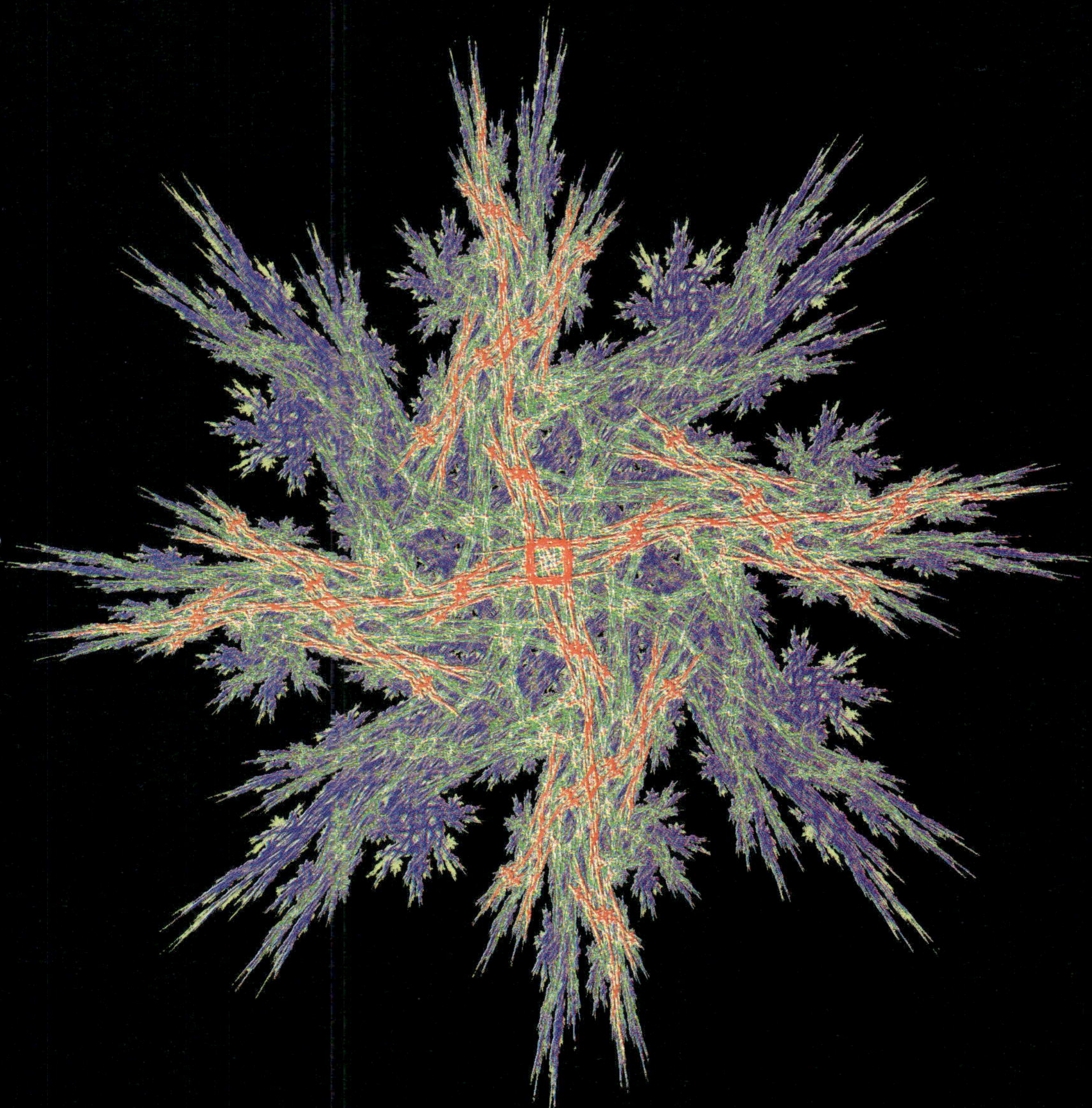

Figure 7.16 (on previous pages 184–5) *Symmetric fractal with (a) symmetric coloring and (b) asymmetric coloring.*

to the vertex A and then rotates it through 120°. Considering the geometry, it is relatively easy to see that this is the same as first rotating X through 120° and then moving halfway to the vertex B. In symbols, $r_{120}\rho_A = \rho_B r_{120}$. Similarly, $r_{240}\rho_A = \rho_C r_{240}$.

Since Z is $\mathbf{Z}_3$ symmetric, we have $r_{120}(Z) = Z$, $r_{240}(Z) = Z$, and so

$$r_{120}\rho_A(Z) = \rho_B r_{120}(Z) = \rho_B(Z),$$
$$r_{240}\rho_A(Z) = \rho_C r_{240}(Z) = \rho_C(Z).$$

But $Z = \rho_A(Z) \cup r_{120}\rho_A(Z) \cup r_{240}\rho_A(Z)$, and so we have shown that Z satisfies (SIER): $Z = \rho_A(Z) \cup \rho_B(Z) \cup \rho_C(Z)$. As we explained above, it follows that Z is the Sierpiński triangle.

Thus far, we have shown that the Sierpiński triangle has $\mathbf{Z}$ symmetry. In fact, the Sierpiński triangle has $\mathbf{D}_3$ symmetry and we use a similar argument to establish this extra symmetry. To see why this is so, let R denote the reflection in the x-axis. We already know that the Sierpiński triangle is equal to the set Z which has $\mathbf{Z}_3$ symmetry. To verify $\mathbf{D}_3$ symmetry, it is enough to show that $R(Z) = Z$. We may easily check that

$$R\rho_A = \rho_A R, \quad R\rho_B = \rho_C R, \quad R\rho_C = \rho_B R$$

It follows that $R\rho_A(Z) = \rho_A R(Z)$, $R\rho_B(Z) = \rho_C R(Z)$, $R\rho_C(Z) = \rho_B R(Z)$, and so

$$R(Z) = \rho_A R(Z) \cup \rho_B R(Z) \cup \rho_C R(Z)$$

We make yet another application of the uniqueness part of the first theorem to deduce that $R(Z) = Z$.

Structure on all scales

Perhaps the most exciting feature of fractals is that they have structure on all scales. Although we will not discuss this issue further

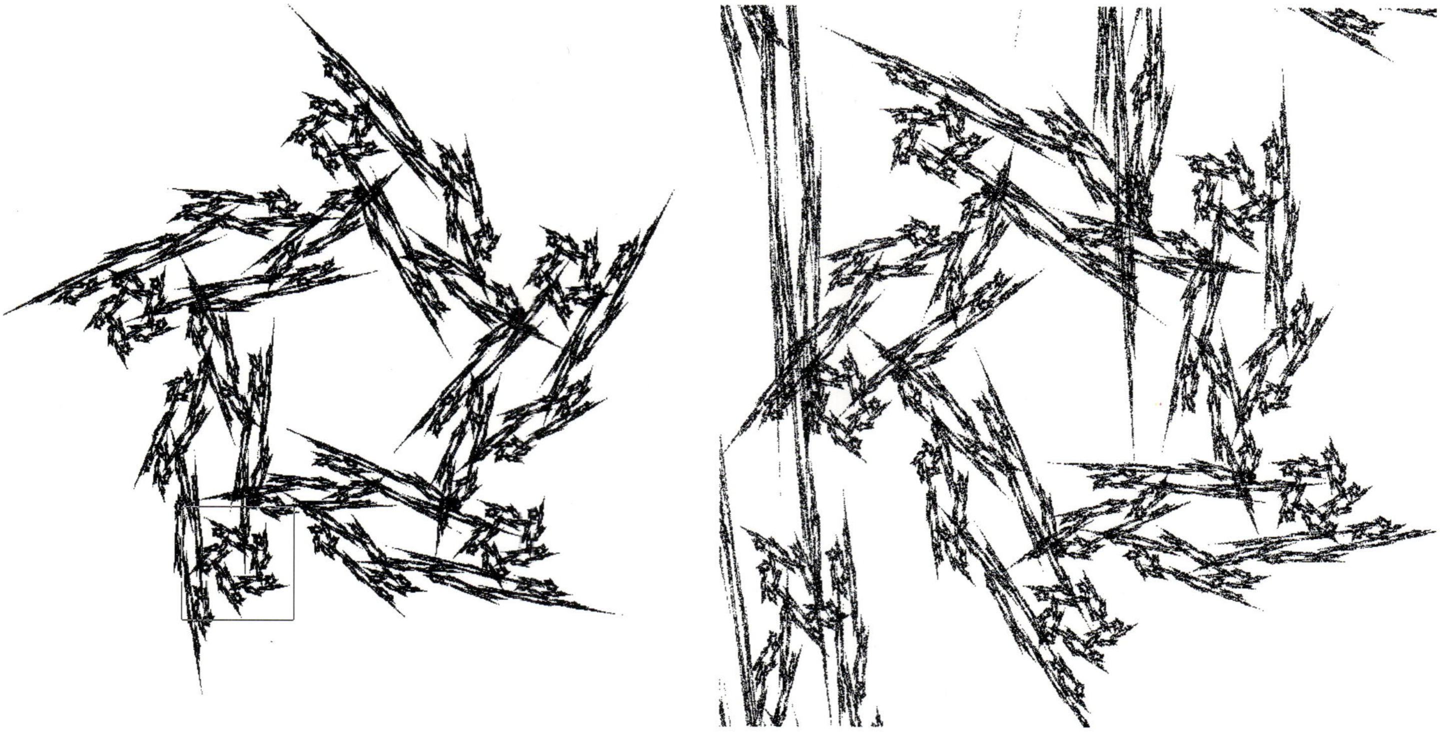

(a) (b)

Figure 7.17 Magnifying a fractal

here, we can illustrate the phenomenon with two pictures (Figure 7.17), the second showing a magnification of part of the first. Observe how the symmetry in Figure 7.17(b) appears approximately in the new scale. In Figure 7.17(a) we have taken a small box around one branch of the fractal and magnified it by about five times in Figure 7.17(b).

More examples

We conclude the chapter with some further observations about the two theorems on iterated function systems. Note that the first theorem giving the existence of symmetric fractals did not depend on ideas of randomness or probability. With this in mind, it is worthwhile thinking about the way we actually constructed fractals using the random selection of contractions from an iterated function system. Suppose, for example, that we had an iterated function system consisting of three affine contractions ρ_1, ρ_2, and ρ_3. Suppose instead of choosing each contraction with equal probability, we decided to choose ρ_1 with probability $\frac{1}{2}$ and ρ_2

and ρ_3 with probability $\frac{1}{4}$. For example, we could choose whole numbers randomly between 1 and 4 and select ρ_1 if the random number was either 1 or 2, ρ_2 if it were 3, and ρ_3 if it were 4. One can show that the process again converges with probability one to the fractal defined by ρ_1, ρ_2, and ρ_3. However, since we will be choosing ρ_1 more often we would expect to get a different coloring of the fractal. We give an illustration of this in Figure 7.16. The symmetric fractal (a) was constructed in the usual way and both the fractal and the coloring have eightfold symmetry. In (b) we carried out the iteration by selecting the second, fourth, sixth, and eighth elements of the iterated function system with twice the probability of the first, third, fifth, and seventh elements. As you can see, the resulting figure is the same as that shown in (a). However, the coloring now has only fourfold symmetry.

Figure 7.18 (opposite) *Cashmere.*

APPENDICES

Appendix A

PICTURE PARAMETERS

In this appendix, we list the mappings and parameter values that were used to produce all of the images of symmetric chaos pictured in this volume. This information appears in the tables below. We also note, where relevant, the names that we have associated with these figures. The pictures divide into three categories: icons, quilts, and fractals. For the icons, we have used two formulas: one which allows us to change the symmetry of the mapping from $\mathbf{D}_n$ to $\mathbf{Z}_n$ and the other which has a non-polynomial term. For the quilts, we have used two formulas: one for the square quilt patterns and one for the hexagonal quilt patterns. The fractals are all created from the choice of one affine linear map.

Symmetric icons

Each of the formulas that we have used to produce symmetric icons has five real parameters and each has an integer parameter n, the degree of symmetry.

The first and most basic formula used in the computation of the symmetric icons is

$$F(z) = [\lambda + \alpha z\bar{z} + \beta\,\mathrm{Re}(z^n) + \omega i]z + \gamma\bar{z}^{n-1},$$

where α, β, γ, λ, and ω are the real parameters. This formula has either $\mathbf{D}_n$ symmetry (when $\omega = 0$) or $\mathbf{Z}_n$ symmetry (when $\omega \neq 0$). The parameter values for the icons produced using this formula are given in Table A.1.

In the second formula, we omit the ω term (so the formula is always $\mathbf{D}_n$ symmetric) and add in a non-polynomial term. The coefficient of this term is δ, so there are still five real parameters. The purpose of the non-polynomial term is to explore what happens to the attractors when a mild singularity is introduced at the origin but with symmetry preserved. The degree of the singularity present is determined by a second integer p. The formula with the non-polynomial term is

$$F(z) = [\lambda + \alpha z\bar{z} + \beta\,\mathrm{Re}(z^n) + \delta\,\mathrm{Re}([z/|z|]^{np})|z|]z + \gamma\bar{z}^{n-1}.$$

The parameter values for the icons computed using this formula are given in Table A.2.

Symmetric quilts

The formula that we use for computing the square quilt designs is

$$\begin{aligned} f(x, y) = {} & m(x, y) + v + \lambda\,(\sin 2\pi x, \sin 2\pi y) \\ & + \alpha(\sin 2\pi x \cos 2\pi y, \sin 2\pi y \cos 2\pi x) \\ & + \beta(\sin 4\pi x, \sin 4\pi y) \\ & + \gamma(\sin 6\pi x \cos 4\pi y, \sin 6\pi y \cos 4\pi x). \\ & - \omega(\sin 2\pi y, \sin 2\pi x). \end{aligned}$$

This formula contains five real parameters λ, α, β, γ, and ω, an integer m, and a shift which is either $v = (0, 0)$ or $v = (\frac{1}{2}, \frac{1}{2})$. This quilt formula is $\mathbf{D}_4$ symmetric when $\omega = 0$ and $\mathbf{Z}_4$ symmetric when $\omega \neq 0$. The parameter values for the square quilt designs are given in Table A.3

Table A1 Data for symmetric icons—standard formula

Figure	λ	α	β	γ	ω	Symmetry	Name
1.1	−2.7	5.0	1.5	1.0	0.0	$\mathbf{D}_6$	Halloween
1.2	−2.08	1.0	−0.1	0.167	0.0	$\mathbf{D}_7$	Mayan Bracelet
1.8	1.56	−1.0	0.1	−0.82	0.12	$\mathbf{Z}_3$	Clam Triple
1.13	−1.806	1.806	0.0	1.0	0.0	$\mathbf{D}_5$	Emperor's Cloak
1.15/1.17	1.56	−1.0	0.1	−0.82	0.0	$\mathbf{D}_3$	The Trampoline
1.18(a)	−2.18	10.0	−12.0	1.0	0.0	$\mathbf{Z}_2$	Fish and Eye
1.18(b)	−2.195	10.0	−12.0	1.0	0.0	$\mathbf{D}_3$	
2.3	−1.86	2.0	0.0	1.0	0.1	$\mathbf{Z}_4$	Swirling Streamers
3.5(b)	−2.34	2.0	0.2	0.1	0.0	$\mathbf{D}_5$	The Sanddollar
3.6(b)	2.6	−2.0	0.0	−0.5	0.0	$\mathbf{D}_5$	Pentagon Attractor
3.7(b)	−2.5	5.0	−1.9	1.0	0.188	$\mathbf{Z}_5$	Chaotic Flower
3.11(a)	2.409	−2.5	0.0	0.9	0.0	$\mathbf{D}_{23}$	Kachina Dolls
3.13(b)	2.409	−2.5	−0.2	0.81	0.0	$\mathbf{D}_{24}$	Santa Chiara Icon
3.14	−2.05	3.0	−16.79	1.0	0.0	$\mathbf{D}_9$	French Glass
3.15(a)	−2.32	2.32	0.0	0.75	0.0	$\mathbf{D}_5$	The Pentacle
5.5	2.5	−2.5	0.0	0.9	0.0	$\mathbf{D}_3$	Golden Flintstone
5.12(a)	1.455	−1.0	0.03	−0.8	0.0	$\mathbf{D}_3$	
5.13(a)	2.39	−2.5	−0.1	0.9	0.0	$\mathbf{D}_{16}$	
5.13(b)	2.39	−2.5	−0.1	0.9	−0.15	$\mathbf{Z}_{16}$	
5.14	1.5	−1.0	0.1	−0.8	0.0	$\mathbf{Z}_2$	
5.15	1.5	−1.0	0.1	−0.805	0.0	$\mathbf{D}_3$	Trinity in Red

Table A2 Data for symmetric icons—non-polynomial term

Figure	λ	α	β	γ	δ	Symmetry	p	Name
1.7	1.5	−1.0	−0.2	−0.75	0.04	$\mathbf{D}_3$	24	Wild Chaos
2.5	−2.5	8.0	−0.7	1.0	−0.9	$\mathbf{D}_9$	0	Lace by Nine
3.4(b)	−2.38	10.0	−12.3	0.75	0.02	$\mathbf{D}_5$	1	Gothic Medalion
3.9(b)	1.0	−2.1	0.0	1.0	1.0	$\mathbf{D}_3$	1	Mercedes-Benz
3.11(b)	−2.225	1.5	−0.014	0.002	−0.02	$\mathbf{D}_{57}$	0	Sunflower
3.15(b)	−2.42	1.0	−0.04	0.14	0.088	$\mathbf{D}_6$	0	Star of David
5.12(b)	1.455	−1.0	0.03	−0.8	−0.025	$\mathbf{D}_3$	0	

Table A3 Data for square quilts

Figure	λ	α	β	γ	ω	Shift	m	Name
2.6	−0.59	0.2	0.1	−0.33	0.0	(0, 0)	2	Emerald Mosaic
2.12	−0.59	0.2	0.1	−0.27	0.0	$(\frac{1}{2}, \frac{1}{2})$	0	Sugar and Spice
3.18(b)	−0.2	−0.1	0.1	−0.25	0.0	(0, 0)	0	Sicilian Tile
3.20(b)	0.25	−0.3	0.2	0.3	0.0	(0, 0)	1	Roses
3.21(b)	−0.28	0.25	0.05	−0.24	0.0	(0, 0)	−1	Wagonwheels
3.23(e)	−0.12	−0.36	0.18	−0.14	0.0	$(\frac{1}{2}, \frac{1}{2})$	1	Victorian Tiles
3.24(a)	0.1	0.2	0.1	0.39	0.0	(0, 0)	−1	Mosque
3.24(b)	−0.589	0.2	0.04	−0.2	0.0	$(\frac{1}{2}, \frac{1}{2})$	0	Red Tiles
3.25(b)	−0.28	0.08	0.45	−0.05	0.0	$(\frac{1}{2}, \frac{1}{2})$	0	Cathedral Attractor
6.1	−0.59	0.2	0.2	0.3	0.0	(0, 0)	2	Gyroscopes
6.2	−0.28	0.25	0.05	−0.24	0.0	$(\frac{1}{2}, \frac{1}{2})$	−1	Cats Eyes
6.3	−0.11	−0.26	0.19	−0.059	0.07	$(\frac{1}{2}, \frac{1}{2})$	2	Flowers with Ribbons

Table A4 Data for hexagonal quilts

Figure	λ	α	β	γ	ω	m	Name
2.9	0.1	−0.076	−0.6	0.1	0.0	0	Dutch Quilt
2.13	0.2	0.04	0.1	0.1	0.0	1	Crown of Thorns
3.28(b)	−0.105	−0.15	0.06	−0.03	0.0	0	Hexagonal Design
6.4	0.02	−0.01	0.14	0.052	0.0	0	Sundial Mosaic
6.5	0.02	−0.1	0.14	0.052	0.04	0	Fractured Symmetry
D.8	0.02	−0.1	0.14	0.052	0.0	$e^{i\pi/3}$	Hex Nuts
D.9	0.1	−0.076	−0.06	0.1	0.101	0	Marching Troupe

The formula that we have used to create hexagonal quilt designs is

$$\begin{aligned} f(\boldsymbol{X}) = m\boldsymbol{X} &+ \Lambda[\sin(2\pi\, \boldsymbol{N}\cdot \boldsymbol{X})\boldsymbol{N} + \sin(2\pi\, R\boldsymbol{N}\cdot \boldsymbol{X})R\boldsymbol{N} \\ &+ \sin(2\pi\, R^2\boldsymbol{N}\cdot \boldsymbol{X})R^2\boldsymbol{N}] + \alpha[\sin(2\pi\, \boldsymbol{M}\cdot \boldsymbol{X})\boldsymbol{M} \\ &+ \sin(2\pi\, R\boldsymbol{M}\cdot \boldsymbol{X})\, R\boldsymbol{M} + \sin(2\pi\, R^2\boldsymbol{M}\cdot \boldsymbol{X})R^2\boldsymbol{M}] \\ &+ \sin(2\pi\, \boldsymbol{L}\cdot \boldsymbol{X})\boldsymbol{a} + \sin(2\pi\, R\boldsymbol{L}\cdot \boldsymbol{X})R\boldsymbol{a} \\ &+ \sin(2\pi\, R^2\boldsymbol{L}\cdot \boldsymbol{X})R^2\boldsymbol{a} + \sin(2\pi\, F\boldsymbol{L}\cdot \boldsymbol{X})F\boldsymbol{a} \\ &+ \sin(2\pi\, RF\boldsymbol{L}\cdot \boldsymbol{X})RF\boldsymbol{a} + \sin(2\pi\, R^2F\boldsymbol{L}\cdot \boldsymbol{X})R^2F\boldsymbol{a}, \end{aligned}$$

where $\boldsymbol{X} = (x, y)$ is a point in the plane, $\boldsymbol{L} = (3, \frac{1}{\sqrt{3}})$, $\boldsymbol{M} = (2, 0)$ and $\boldsymbol{N} = (1, -\frac{1}{\sqrt{3}})$. The symbols R, F, and Λ denote linear transformations of the plane; R is rotation of the plane counter-clockwise by 120°, and F is reflection in the x-axis, and Λ corresponds to complex multiplication by $\lambda + i\omega$. We have used the dot product notation $\boldsymbol{N}\cdot\boldsymbol{X} = n_1x + n_2y$, where $\boldsymbol{N} = (n_1, n_2)$ and $\boldsymbol{X} = (x, y)$.

This formula also has five real parameters λ, α, $\boldsymbol{a} = (\beta, \gamma)$, and ω; it has $\mathbf{D}_6$ symmetry when $\omega = 0$ and $\mathbf{Z}_6$ symmetry when $\omega \neq 0$. The parameter m is usually an integer, but it can take values which are integer multiples of a sixth root of unity. When m is not an integer, the resulting figure will have only $\mathbf{Z}_6$ symmetry. Parameter values for the hexagonal quilt patterns that we have shown are given in Table A.4.

Symmetric fractals

Finally, we consider the symmetric fractals. Each symmetric fractal is created from one affine linear mapping and an affine linear mapping is determined by six real constants: a 2 × 2

Table A5 Data for symmetric fractals

Figure	a_{11}	a_{12}	a_{21}	a_{22}	b_1	b_2	Symmetry	Name
7.1	−0.4	0.75	0.2	−0.3	0.0	0.4	$\mathbf{Z}_{55}$	Whipper-snipper
7.3	0.5	0.0	0.0	0.5	0.5	0.0	$\mathbf{Z}_3$	Sierpiński Triangle
7.4(a)	0.5	0.0	0.0	0.5	0.5	0.0	$\mathbf{Z}_5$	Sierpiński Pentagon
7.4(b)	0.5	0.0	0.0	0.5	0.5	0.0	$\mathbf{Z}_6$	Sierpiński Hexagon
7.7(a)	0.45	0.0	0.0	0.45	0.55	0.0	$\mathbf{Z}_3$	Strict fractal
7.7(c)	0.55	0.0	0.0	0.55	0.45	0.0	$\mathbf{Z}_3$	Overlaid fractal
7.8	0.45	−0.1	−0.31	0.45	0.1	0.2	$\mathbf{Z}_{11}$	Circular Saw
7.9	0.4	−0.1	−0.35	0.4	0.01	0.2	$\mathbf{Z}_9$	Catherine Wheel
7.10	−0.1	0.35	0.2	0.5	0.5	0.4	$\mathbf{D}_3$	Bee
7.12	0.46	0.0	0.0	0.46	0.54	0.0	$\mathbf{Z}_4$	Sierpiński Square Variation
7.13	−0.25	−0.3	0.3	−0.26	0.5	0.5	$\mathbf{D}_8$	Doily
7.14	−0.25	−0.3	0.3	−0.34	0.5	0.5	$\mathbf{D}_4$	Astigmatism
7.15	−0.25	−0.3	0.14	−0.26	0.5	0.5	$\mathbf{Z}_{12}$	Paisley
7.16	0.45	−0.1	0.3	−0.4	0.15	0.1	$\mathbf{Z}_8$	Asymmetric coloring
7.17	0.4	−0.1	−0.31	0.45	0.01	0.0	$\mathbf{Z}_5$	Magnifying a fractal
7.18	−0.15	0.75	0.2	−0.3	0.075	0.4	$\mathbf{Z}_{50}$	Cashmere

matrix consisting of four constants a_{11}, a_{21}, a_{12}, and a_{22}, and a translation vector consisting of two real constants b_1 and b_2. The symmetric fractals are created from the affine linear mapping by choosing randomly from the elements of a finite group, in this case either $\mathbf{D}_n$ or $\mathbf{Z}_n$. The data that we used to produce the various symmetric fractals are listed in Table A.5.

Appendix B

BASIC PROGRAMS

In this appendix, we present computer programs written in Basic that will produce symmetric icons, symmetric fractals, quilts on a square lattice, and quilts on a hexagonal lattice. Primitive coloring schemes are given for the quilt patterns. These programs are written in Microsoft QuickBasic, but can be converted easily to standard Basic. (Line numbers must be added for all lines, and alphabetic labels, such as *loops* and *iterate*, must be replaced by the corresponding line numbers.)

It goes without saying that the faster your computer is, the more enjoyable you will find playing with these programs. Decent speed can be obtained on a 80286 computer with a mathematics coprocessor (80287 chip). The programs will work with sufficient speed on an 8086 machine with a mathematics coprocessor (8087 chip) or on a 80286 computer without a mathematics coprocessor. Since the programs are computationally intensive, you will find them painfully slow on a 8086 computer without a mathematics coprocessor. If you have access to a Basic compiler, then a significant increase in speed can be obtained by compiling the programs. The symmetric fractal program is the speediest, with the symmetric icon program a close second. Don't even try the quilt programs, which involve many evaluations of *sine* and *cosine* on each iteration, without a mathematics coprocessor.

Each program contains:

(1) a loop for computing the dynamics;

(2) a method for graphing the results;

(3) a menu for changing the parameters in the mapping;

(4) an initialization section;

(5) a primitive error trap routine.

The mathematical heart of each program is the iteration routine included in the section from *loops:* to *GOTO loops*. This section of the program creates an intentional infinite loop where the desired function is iterated as often as one likes and the results graphed. There is one escape from this loop. By pressing *m*, the program exits from the loop and brings up the menu, which allows further instructions to be entered. While in this loop, entering *c* clears the screen and then *continues* the iteration from the point where the screen is cleared. Entering *i* puts the current iterate number in the upper left hand corner of the screen.

The actual function evaluation is done in the section of the program entitled *iterate:*. The functions that are included here are the ones indicated in the previous chapters. The most difficult part of each program is the graphics section. The actual graphing command is the *PSET* command in the *loops:* section. In Basic, one must tell the computer what kind of graphics screen the computer has. We have assumed that the computer is an IBM compatible with VGA graphics. This information is entered in the *initialize:* subroutine. The constant *nscreen* = 12 indicates VGA graphics. The value of *nscreen* must be changed if your computer has another graphics adapter. For example, *nscreen* = 1 is used for CGA graphics. The program also needs to know the number of pixels on your computer screen: the number in the horizontal direction is called *npixelx* and the number in the vertical direction is *npixely*. These numbers are 640 and 480 for VGA graphics and 320 and 200 for CGA graphics and are set in the *initialize:* subprogram. The graphics part of these programs divide the computer screen into two parts (see Figure B.1). The current parameter values for the mapping are displayed in the left part while the actual results of the iteration process are displayed on the right. A vertical line is drawn dividing the two parts. The pixel coordinate where this vertical line is drawn must be set in the *initialize:* subprogram. The variable to be set is *nstartx* and this variable should be set to 160 for VGA graphics and 60 for CGA graphics. Some experimentation may be needed for other graphics adapters.

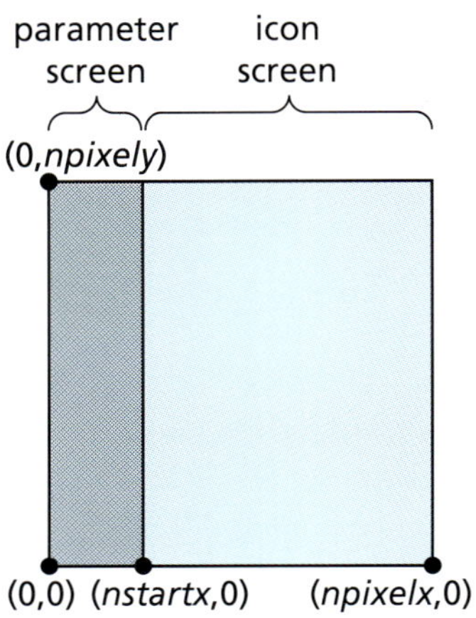

Figure B.1 *Computer screen parameters.*

The functions *DEF fnxpix* and *DEF fnypix* converts planar coordinates to pixel coordinates and it is the pixel coordinates that are lit by the *PSET* command. In the symmetric icon and symmetric fractal programs, there is also a parameter called scale. We ask the computer to plot only those points that fall between $-scale$ and *scale* in both the real x and y coordinates. In the *menu:* subroutine, there is a way to change *scale* when you find that the iteration process is taking you out of the graphics window. Finally, it is possible (and indeed not unusual in the symmetric icon program) that the iteration process will blow up, that is, iterates will march off unboundedly to infinity. The program is equipped with a primitive error trap (*ON ERROR GOTO errortrap*) to handle this eventuality. When numerical overflow (or for that matter any other error) occurs, the *menu:* subprogram is automatically called and the initial conditions are reset to fixed small values.

Each of the programs is equipped with a *menu:* subprogram between menu: and *GOTO menu*. The menu works as follows. When the menu appears on the screen, if you type the letter that is capitalized, then, with one exception, the computer executes the associated command. The exception is the *Escape* key, which is used to exit the program. For example, to change the parameter *alpha*, type *a* and the computer will ask you for a new value of *alpha*, which you then enter followed by pressing the *Return* key. There are two important keys in the *menu: r* begins the iteration process with the parameter values that have been set, and *x* puts you in a little subprogram that resets the initial conditions, either automatically by entering r or manually by entering *x*. If manual input is chosen, then one enters the two coordinates *x*, *y* (with the comma). IMPORTANT NOTICE: the menu accepts entries only in LOWER CASE!

Symmetric Icon Program: Program1

```
          DEFDBL I, P-Q, X-Z
          ON ERROR GOTO errortrap
          DEF fnxpix (x) = nstartx + scalex * (x + scale)
          DEF fnypix (y) = npixely – scalex * (y + scale)
          GOSUB initialize
          GOSUB menu
loops:
          GOSUB iterate
          x = xnew: y = ynew
          PSET (fnxpix(x), fnypix(y))
          a$ = INKEY$
          IF a$ = "c" THEN iterates = 1: CLS: GOSUB parameters
          IF a$ = "i" THEN GOSUB parameters
restart:
          IF a$ = "m" THEN GOSUB menu
          iterates = iterates + 1
          GOTO loops
iterate:
          zzbar = x * x + y * y
          zreal = x: zimag = y
          FOR i = 1 TO n – 2
          za = zreal * x – zimag * y
          zb = zimag * x + zreal * y
          zreal = za: zimag = zb
          NEXT i
          zn = x * zreal – y * zimag
          p = lambda + alpha * zzbar + beta * zn
          xnew = p * x + gamma * zreal – omega * y
          ynew = p * y – gamma * zimag + omega * x
          RETURN
menu:
          GOSUB parameters
          PRINT USING "(X,y) = ##.#### ##.####"; x; y
          PRINT "Scale =", scale
          PRINT " ESC to exit program"
          PRINT " R to return to iteration"
1010:
          b$ = INKEY$
          iterates = 1
          IF b$ = "" THEN 1010
```

```
        IF b$ = "l" THEN INPUT "lambda = ", lambda
        IF b$ = "a" THEN INPUT "alpha =", alpha
        IF b$ = "b" THEN INPUT "beta =", beta
        IF b$ = "g" THEN INPUT "gamma =", gamma
        IF b$ = "o" THEN INPUT "omega =", omega
        IF b$ = "d" THEN INPUT "degree of symmetry =", n
        IF b$ = "x" THEN GOSUB initialpoint
        IF b$ = "r" THEN iterates = 1: GOSUB parameters: RETURN
        IF b$ = "s" THEN INPUT "scale =", scale: GOSUB setscreen
        IF b$ = CHR$(27) THEN STOP
        CLS
        GOTO menu
parameters:
        LOCATE 1, 1
        PRINT "iterates =", iterates
        PRINT "Lambda =", lambda
        PRINT "Alpha =", alpha
        PRINT "Beta  =", beta
        PRINT "Gamma =", gamma
        PRINT "Omega =", omega
        PRINT "Degree =", n
        LINE (nstartx, 0)-(nstartx, npixely)
        RETURN
initialize:
        scale = 1
        nscreen = 12: npixelx = 640: npixely = 480
        nstartx = 160
        SCREEN nscreen
        GOSUB setscreen
        x = .01: y = .003: n = 4: iterates = 1
        lambda = -1.8: alpha = 2: beta = 0: gamma = 1: omega = 0
        RETURN
errortrap:
        x = .0234: y = .12345
        a$ = "m"
        RESUME restart
initialpoint:
        CLS
        PRINT "Enter r to reset coordinates automatically"
        PRINT "Enter x to INPUT coordinates"
3010:
        c$ = INKEY$
        IF c$<> "r" AND c$<> "x" THEN 3010
        IF c$ = "r" AND c$"x"THEN x = .003: y = .0005345
        IF c$ = "x" THEN INPUT "(x,y) coordinates =", x, y
        xnew = x: ynew = y
        RETURN
setscreen:
        CLS
        scaley = npixely / (2 * scale)
        scalex = (npixelx – nstartx) / (2 * scale)
        RETURN
```

Symmetric Fractals Program

The features in the symmetric fractals program are similar to those in the symmetric icons program with two exceptions. First, the affine map on which the iterated function system is based is written in the form

$$\rho(x, y) = \begin{pmatrix} a_{11}x + a_{12}y + b_1 \\ a_{21}x + a_{22}y + b_2 \end{pmatrix}.$$

In the *menu*, each of the six constants has a number from 1 to 6 preceding it (e.g. 1. a_{11}); to change the value of that constant enter that number (e.g. enter 1 to change a_{11}).

Second, the dynamics of the iterated function system will converge only when the affine map ρ is a contraction, that is, when

$$a_{11}^2 + a_{21}^2 > 1,$$
$$a_{21}^2 + a_{22}^2 > 1,$$
$$a_{11}^2 + a_{21}^2 + a_{21}^2 + a_{22}^2 > 1 + (a_{11}a_{22} - a_{21}a_{12})^2.$$

For this reason, a warning message will appear when parameters are entered for which ρ is not a contraction.

Symmetric Fractal Program: Program 2

```
        DEFDBL I, X-Z
        DEFINT M-N
        DIM c(100), s(100)
        ON ERROR GOTO errortrap
        DEF fnxpix (x) = nstartx + scalex * (x + scale)
        DEF fnypix (y) = npixely – scalex * (y + scale)
        GOSUB initialize
        GOSUB menu
loops: Same as Program 1 for symmetric icons
restart: Same as Program 1 for symmetric icons
iterate:
        xnew = a11 * x + a12 * y + b1
        ynew = a21 * x + a22 * y + b2
        m = INT(n * RND)
        x1 = xnew: y1 = ynew
        xnew = c(m) * x1 – s(m) * y1
        ynew = s(m) * x1 + c(m) * y1
        IF conj = 0 THEN RETURN
        m = INT(2 * RND)
        IF m = 1 THEN ynew = -ynew
        RETURN
```

```
menu:
        GOSUB parameters
        PRINT "Degree of symmetry "; n
        PRINT USING "(X,y) = ##.#### ##.####"; x; y
        PRINT "Scale = "; scale
        IF conj = 0 THEN PRINT "Toggle for Dn symmetry"
        IF conj = 1 THEN PRINT "Toggle for Zn symmetry"
        PRINT "ESC to exit program"
        PRINT "R to return to iteration"
1010:
        b$ = INKEY$
        iterates = 1
        IF b$ = "" THEN 1010
        IF b$ = CHR$(27) THEN STOP
        IF b$ = "d" THEN INPUT "degree of symmetry = ", n:GOSUB trig
        IF b$ = "t" THEN conj = 1 – conj
        IF b$ = "1" THEN INPUT "new a11 = ", a11
        IF b$ = "2" THEN INPUT "new a12 = ", a12
        IF b$ = "3" THEN INPUT "new a21 = ", a21
        IF b$ = "4" THEN INPUT "new a22 = ", a22
        IF b$ = "5" THEN INPUT "new b1 = ", b1
        IF b$ = "6" THEN INPUT "new b2 = ", b2
        IF b$ = "x" THEN GOSUB initialpoint
        IF b$ = "r" THEN iterates = 1: GOSUB parameters RETURN
        IF b$ = "s" THEN INPUT "scale = ", scale
        GOSUB setscreen
        CLS
        GOTO menu
parameters:
        LOCATE 1, 1
        PRINT "iterates = ", iterates
        PRINT USING "symmetry = !##"; d$(conj); n
        PRINT "1. a11 = "; a11
        PRINT "2. a12 = "; a12
        PRINT "3. a21 = "; a21
        PRINT "4. a22 = "; a22
        PRINT "5. b1 = "; b1
        PRINT "6. b2 = "; b2
        a1 = a11 * a11 + a21 * a21: a2 = a21 * a21 + a22 * a22
        IF a1 > 1 OR a2 > 1 OR a1 + a2 > 1 + (a11 * a22 —a12 * a21)^2
        THEN PRINT "WARNING - Affine mapping is NOT a contraction"
        LINE (nstartx, 0)-(nstartx, npixely)
        RETURN
initialize:
        CLS
        scale = 1
        nscreen = 12: npixelx = 640: npixely = 480
        nstartx = 160
        SCREEN nscreen
        GOSUB setscreen
        a11 = .4: a12 = .35: a21 = .2: a22 = .4
        b1 = 0: b2 = .4: iterates = 1
        x = .1: y = -.01: n = 3: conj = 1
        pi = 355 / 113
        GOSUB trig
        d$(0) = "Z": d$(1) = "D"
        RETURN
errortrap: Same as Program 1 forsymmetric icons
initialpoint: Same as Program 1 forsymmetric icons
setscreen: Same as Program 1 forsymmetric icons
trig:
        FOR i = 0 TO n – 1
        c(i) = COS(2 * pi * i / n)
        s(i) = SIN(2 * pi * i / n)
        NEXT i
        RETURN
```

Square Quilts Program

The basic iteration process for the square quilts program is similar to that of the symmetric icon program. Mathematically, the only essential difference is found in the *iterate:* subprogram where the fact that iteration is restricted to the unit square is imposed. The four *IF* statements force the new x and y values to lie in the unit square. The iteration formula was presented in Chapter 6 and Appendix A and is derived in Appendix D. Recall that the shift $\mathbf{v}$ may either be $(0, 0)$ or $(\frac{1}{2}, \frac{1}{2})$. In the *menu* pressing *h* toggles between the two possibilities.

On the programming level, the major differences between this program and the previous ones involve graphics and it is here that there are two differences: periodicity and color. Since the attractors on the unit square can be reproduced periodically, we provide a new control parameter *nperiod* in the *menu* that gives the number of times in *both* the horizontal and vertical directions that the attractor in the unit square is to be repeated. To begin, *nperiod* is set to 3; thus the basic pattern on the unit square is reproduced nine times in a 3×3 array. This complicates the plotting routine, so, instead of just including it in the *loops:* subprogram as we did previously, we introduce a new subroutine labelled *plotpoints:*.

Major programming changes are also due to the addition of a primitive coloring routine. We have indicated statements that are included only because of coloring by a *C*; these statements may be omitted if the coloring feature is *not* desired. If the coloring feature is included, you must *delete* the *C* when typing those lines. When coloring is included, we allow the program to run either with or without coloring. Pressing *t* in the *menu* toggles between these options.

We now describe how the coloring feature works. The major point is that now we must count how many times each pixel in the unit square is hit during the iteration process, and we must have an integer array to store this data. The storage issue places a major strain on Basic's capabilities. Roughly speaking the largest array of integers that a Basic program can store is 160 × 160. This array is not large enough to store data for the full screen; even in CGA one needs a 200 × 200 array. For the quilts, however, we need only store data for coloring on the unit square. Thus, if we are trying to color an $N \times N$ array of pixels, we need only store data for an $(N/nperiod \times N/nperiod)$ array. For example, in VGA graphics, the pixel array we are trying to color is 480 ×480. As long as *nperiod* is at least 3, Basic has the capability of storing the needed data, since $480/nperiod \leq 160$. Finally we discuss the details of the coloring program. We assume that the number of (non-black) colors available is 15 (*ncolor* = 15 in the *initialization:* subprogram). The colors 1–15 are ordered in the *setcolors:* subprogram. (Depending on how your version of Basic orders colors, you may want to change this order.) For easy reference the *setscreen:* subprogram displays the colors in the menu part of the screen.

The mechanics of the coloring works as follows. Given the number of times a pixel has been hit, *mc*, we color that pixel in the following way. If

$$\begin{array}{lll} m_1 = 1 & \leq mc < m_2 & \text{choose color 1,} \\ m_2 & \leq mc < m_3 & \text{choose color 2,} \\ \vdots & & \vdots \\ m_{14} & \leq mc < m_{15} & \text{choose color 14,} \\ m_{15} & \leq mc & \text{choose color 15.} \end{array}$$

The method by which we accomplish this scheme is to create an array *mcolor* such that

$$mcolor(1) = \cdots = mcolor(m_2 - 1) = colord(1),$$
$$mcolor(m_2) = \cdots = mcolor(m_3 - 1) = colord(2),$$

and so on. This array is also set in the *setcolors:* subprogram. For simplicity we have set $m_j = j$. Thus the color of a pixel changes each time it is hit, up to 15 times (which is the maximum number of colors allowed). The color of that pixel then remains at the 15th color, no matter how many times it is hit. Unless your computer can perform large numbers of iterates very quickly, you will rarely want to change this coloring routine, but the option for change is included. Recall that in order to obtain the high-resolution pictures we have shown in this book, we had to perform millions of iterations. Realistically, one will not be able to compute more than 100,000 iterates in any reasonable length of time on an IBM 80286 clone with a mathematics coprocessor.

Square Quilts Program: Program 3

```
        DEFDBL I, P-Q, X-Z
        DEFINT M
        ON ERROR GOTO errortrap
C         DIM mcount(160, 160), mcolor(200),colord(256)
        DEF fnxpix (x) = nstartx + (npixelx – nstartx) * x / nperiod
        DEF fnypix (y) = npixely - npixely * y / nperiod
        GOSUB initialize C    GOSUB setcolors
        GOSUB menu
        iterates = 1
loops:
        GOSUB iterate
        x = xnew: y = ynew
        GOSUB plotpoints
        a$ = INKEY$
        IF a$ = "c" THEN iterates = 1: CLS: GOSUB setscreen
        IF a$ = "i" THEN LOCATE 1, 1: PRINT "iterates ="; iterates
restart:
        IF a$ = "m" THEN GOSUB menu
        iterates = iterates + 1
        GOTO loops
iterate:
        sx = SIN(p2 * x): sy = SIN(p2 * y)
        xnew = (lambda + alpha * COS(p2 * y)) * sx - omega * sy +beta *
        SIN(2 * p2 * x) + gamma * SIN(3 * p2 * x) * COS(2 * p2 * y)+ma *
        x + shift
        ynew = (lambda + alpha * COS(p2 * x)) * sy + omega * sx +beta *
        SIN(2 * p2 * y) + gamma * SIN(3 * p2 * y) * COS(2 * p2 * x)+ma *
        y + shift
        IF xnew > 1 THEN xnew = xnew – INT(xnew)
        IF ynew > 1 THEN ynew = ynew – INT(ynew)
        IF xnew < 0 THEN xnew = xnew + INT(-xnew) + 1
        IF ynew < 0 THEN ynew = ynew + INT(-ynew) + 1
C         IF toggle = 0 THEN RETURN
C         mxnew = xnew * (npixelx – nstartx) / nperiod: mynew = ynew *
          npixely / nperiod
C         mcount(mxnew, mynew) =mcount(mxnew, mynew) + 1
        RETURN menu:
        GOSUB parameters
        PRINT "No. periods = "; nperiod
        PRINT USING "(X,y)= ##.#### ##.####"; x; y
C         PRINT "T to toggle coloring:";
C         IF toggle = 0 THEN PRINT "coloring off" ELSE PRINT
          "coloring on"
        PRINT "E to exit"
        PRINT "R for RETURN"
1010:
          b$ = INKEY$
```

```
            iterates = 1
            IF b$ = "" THEN 1010
            IF b$ = "l" THEN INPUT "lambda = ", lambda
            IF b$ = "a" THEN INPUT "alpha =", alpha
            IF b$ = "b" THEN INPUT "beta =", beta
            IF b$ = "g" THEN INPUT "gamma =", gamma
            IF b$ = "o" THEN INPUT "omega =", omega
            IF b$ = "m" THEN INPUT "m =", ma
C           IF b$ = "t" THEN toggle = 1 – toggle
            IF b$ = "h" THEN shift = .5 – shift
            IF b$ = "n" THEN INPUT "# of periods =", nperiod
            IF b$ = "x" THEN GOSUB initialpoint
            IF b$ = "r" THEN iterates = 1: GOSUB setscreen: RETURN
            IF b$ = CHR/(27) THEN STOP
            CLS
            GOTO menu
parameters:
            LOCATE 1, 1
            PRINT "iterates= "; iterates
            PRINT "Lambda= "; lambda
            PRINT "Alpha = "; alpha
            PRINT "Beta = "; beta
            PRINT "Gamma = "; gamma
            PRINT "Omega = "; omega
            PRINT "M ="; ma
            PRINT "sHift by "; shift
            LINE (nstartx, 0)-(nstartx, npixely)
            RETURN
initialize:
C           toggle = 0: ncolor = 15
            nperiod = 3
            nscreen = 12: npixelx = 640: npixely = 480
            nstartx = 160
            CLS
            SCREEN nscreen
            pi = 355 / 113: p2 = 2 * pi
            x = .1: y = .334: xnew = x: ynew = y
            lambda = -.59: alpha = .2: beta = .1: gamma = -.09: omega= 0
            ma = 0
            shift = 0
            RETURN
errortrap: Same as Program 1 for symmetric icons
initialpoint: Same as Program 1 for symmetric icons
setscreen:
            CLS
            GOSUB parameters
C             IF toggle = 0 THEN RETURN
C             FOR j = 0 TO (npixelx – nstartx) / nperiod
C             FOR i = 0 TO npixely / nperiod
C             mcount(j, i) = 0
C             NEXT i
C             NEXT j
C             FOR i = 1 TO ncolor
C             LINE (0, npixely – 20 *i)-(15, npixely- 20 * (i + 1)), colord(i), BF
C             NEXT i
            RETURN
C setcolors:
C             colord(1) = 8: colord(2) = 6:colord(3) = 1:colord(4) = 9
C             colord(5) = 3: colord(6) = 11:colord(7) = 2:colord(8) = 10
C             colord(9) = 5: colord(10) = 13:colord(11) = 4:colord(12) = 12
C             colord(13) = 14: colord(14) = 7:colord(15) = 15
C             FOR j = 1 TO 15
C             mcolor(j) = colord(j)
C             NEXT j
C             RETURN
plotpoints:
            FOR i = 0 TO nperiod – 1
            FOR j = 0 TO nperiod – 1
C             IF toggle = 0 THEN PSET (fnxpix(x+ i), fnypix(y + j)): GOTO 160
C             mc = mcount(mxnew,mynew)
C             IF mc < 15 THEN mm = mcolor(mc) ELSE mm = 15
            PSET (fnxpix(x + i), fnypix(y + j)), mm
'REMARK If coloring is not being used, delete ', mm 'from the previous line
160:
            NEXT j
            NEXT i
            RETURN
```

Hexagonal Quilts Program

The hexagonal quilts program works in a fashion similar to the square quilts program. In particular, the coloring routine and the iteration procedure are identical in principle. We indicate the differences. The major difference is that the fundamental cell of the lattice is a parallelogram with an angle of 60° between the two basic vectors, rather than the unit square. This causes various indexing problems in the program that are solved using elementary linear algebra, the details of which we will not discuss here. Similarly, the actual formulas used for the iteration algorithm on the hexagonal lattice are much more complicated than those needed for the square lattice. These 'facts of life' are responsible for the *setvectors:* and *vector3:* subprograms, where the basis vectors for $\mathcal{L}$ and $\mathcal{L}^*$ are set together with the dual lattice vectors that generate the terms in the iteration algorithm, and for the increased length of the *iterate:* subprogram.

Hexagonal Quilts Program: Program 4

Program 4 for hexagonal quilts begins exactly the same as Program 3 for square quilts. The first changes occur in the iterate: *subroutine.*

```
iterate:
        s11 = SIN(p2 * (el11 * x + el12 * y))
        s12 = SIN(p2 * (el21 * x + el22 * y))
        s13 = SIN(p2 * (el31 * x + el32 * y))
        s21 = SIN(p2 * (em11 * x + em12 * y))
        s22 = SIN(p2 * (em21 * x + em22 * y))
        s23 = SIN(p2 * (em31 * x + em32 * y))
        s31 = SIN(p2 * (en11 * x + en12 * y))
        s32 = SIN(p2 * (en21 * x + en22 * y))
        s33 = SIN(p2 * (en31 * x + en32 * y))
        s3h1 = SIN(p2 * (enh11 * x + enh12 * y))
        s3h2 = SIN(p2 * (enh21 * x + enh22 * y))
        s3h3 = SIN(p2 * (enh31 * x + enh32 * y))
        sx = el11 * s11 + el21 * s12 + el31 * s13
        sy = el12 * s11 + el22 * s12 + el32 * s13
        xnew = ma * x + lambda * sx - omega * sy
        ynew = ma * y + lambda * sy + omega * sx
        xnew = xnew + alpha * (em11 * s21 + em21 * s22 + em31 *s23)
        ynew = ynew + alpha * (em12 * s21 + em22 * s22 + em32 *s23)
        xnew = xnew + a11 * s31 + a21 * s32 + a31 * s33
        ynew = ynew + a12 * s31 + a22 * s32 + a32 * s33
        xnew = xnew + ah11 * s3h1 + ah21 * s3h2 + ah31 * s3h3
        ynew = ynew + ah12 * s3h1 + ah22 * s3h2 + ah32 * s3h3
        by = 2 * ynew / sq3: bx = xnew – by / 2
        IF bx > 1 THEN bx = bx – INT(bx)
        IF by > 1 THEN by = by – INT(by)
        IF bx < 0 THEN bx = bx + INT(-bx) + 1
        IF by < 0 THEN by = by + INT(-by) + 1
        xnew = bx * k11 + by * k21: ynew = bx * k12 – by * k22
        IF toggle = 0 THEN RETURN
        mxnew = bx * (npixelx – nstartx) / nperiod:mynew = by * npixely
        / nperiod
        mcount(mxnew, mynew) = mcount(mxnew, mynew) + 1
        RETURN
```

menu: *The* menu*: subprogram is the same as the one in Program 3 for square quilts with two exceptions. The line:*

```
        IF b$ = "h" THEN shift = .5 – shift
```

is deleted while the line

```
        IF b$ = "r" THEN iterates = 1: GOSUB setscreen
        RETURN
```

is changed to:

```
        IF b$ = "r" THEN iterates = 1: GOSUB setscreen: GOSUB vector3
        RETURN
```

parameters: *The same as in Program3 for square quilts.*

initialize: *The same as in Program3 for square quilts until the line after* pi *is defined.*

```
        pi = 355 / 113: p2 = 2 * pi
        sq3 = SQR(3)
        x = .1: y = .3: xnew = x: ynew = y
        gamma = .1: lambda = -.1: alpha = -.076: beta = 0: omega =0: ma = 0
        GOSUB setvectors
        GOSUB vector3
        RETURN
```

errortrap: *The same as in Program 1 for symmetric icons.*

initialpoint: *The same as in Program 1 for symmetric icons.*

setscreen: *The same as in Program 3 for square quilts.*

setcolors: *The same as in Program 3 for square quilts.*

```
plotpoints:
        FOR i = -nperiod / 2 – 1 TO nperiod / 2 + 1
        FOR j = 0 TO nperiod
        xx = fnxpix(x + i * k11 + j * k21)
        IF xx < nstartx THEN 160
C        IF toggle = 0 THEN PSET(xx, fnypix(y +i * k12 + j * k22))
        GOTO 160
C        mc = mcount(mxnew,mynew)
C        IF mc < 15 THEN mm =mcolor(mc) ELSE mm = 15
        PSET (xx, fnypix(y + i * k12 + j * k22)), mm
'REMARK If coloring is not being used, delete ', mm' from the previous line
160:
        NEXT j
        NEXT i
        RETURN
setvectors:
        k11 = 1: k12 = 0
        k21 = 1 / 2: k22 = sq3 / 2
        el11 = 1: el12 = -1 / sq3
        el21 = 0: el22 = 2 / sq3
        el31 = -(el11 + el21): el32 = -(el12 + el22)
        em11 = 2 * el11 + el21: em12 = 2 * el12 + el22
        em21 = 2 * el21 + el31: em22 = 2 * el22 + el32
        em31 = 2 * el31 + el11: em32 = 2 * el32 + el12
        en11 = 3 * el11 + 2 * el21: en12 = 3 * el12 + 2 * el22
        en21 = 3 * el21 + 2 * el31: en22 = 3 * el22 + 2 * el32
        en31 = 3 * el31 + 2 * el11: en32 = 3 * el32 + 2 * el12
        enh11 = 3 * el11 + el21: enh12 = 3 * el12 + el22
        enh21 = 3 * el21 + el31: enh22 = 3 * el22 + el32
        enh31 = 3 * el31 + el11: enh32 = 3 * el32 + el12
        RETURN
vector3:
        a11 = beta: a12 = gamma
        a21 = (-a11 – sq3 * a12) / 2: a22 = (sq3 * a11 – a12) /2
        a31 = -a11 – a21: a32 = -a12 – a22
        ah11 = a11: ah12 = -a12
        ah21 = (-ah11 – sq3 * ah12) / 2: ah22 = (sq3 * ah11 —ah12) / 2
        ah31 = -ah11 – ah21: ah32 = -ah12 – ah22
        RETURN
```

Appendix C

ICON MAPPINGS

How does one actually find mappings on the plane with $\mathbf{D}_n$ symmetry? The answer to this question resides in a field called *invariant theory*. Fortunately, although many of the questions in this theory are devilishly difficult to answer, the answer to the question we ask is relatively simple to explain. We begin with the simplest example and then state the abstract strategy. First we need some basic definitions.

Let G be a group of linear transformations of $\mathbf{R}^m$. For example, G might be the dihedral group $\mathbf{D}_n$ of symmetries of the regular n-sided polygon acting on the plane $\mathbf{R}^2$. Suppose that X is any point of $\mathbf{R}^m$. The *G-orbit* of X is the set of points in $\mathbf{R}^m$ that result from applying all the transformations of G to X. We say that a function on $\mathbf{R}^m$ is *G-invariant* if it is constant on G-orbits. More precisely, if $p : \mathbf{R}^m \to \mathbf{R}$, then p is G-invariant if $p(gX) = p(X)$ for all $g \in G$. A mapping $f : \mathbf{R}^m \to \mathbf{R}^m$ *commutes* with G (or is *G-equivariant*) if $f(gX) = gf(X)$ for all $g \in G$.

Symmetry on the line

We shall look at a simple group of symmetries of the line $\mathbf{R}$. For our group, we take $G = \mathbf{Z}_2$, where G acts as multiplication by ± 1 on $\mathbf{R}$. In this instance, a function $p : \mathbf{R} \to \mathbf{R}$ will be G-invariant if $p(-X) = p(X)$. Such functions p are called *even* functions. The name 'even' is not plucked from thin air. For should $p(X)$ be a polynomial, that is,

$$p(X) = a_0 + a_1X + a_2X^2 + \cdots + a_mX^m,$$

then

$$p(-X) = a_0 - a_1X + a_2X_2 - \cdots + (-1)^m a_m X^m.$$

It follows that the identity $p(-X) = p(X)$ holds for all X precisely when all of the odd terms vanish. That is,

$$a_1 = a_3 = a_5 = \cdots = 0.$$

Thus

$$p(X) = a_0 + a_2X^2 + \cdots + a_{2k}X^{2k}.$$

A rather simple observation concerning even polynomials is

$$p(X) = q(U),$$

where $U = X^2$ and

$$q(U) = a_0 + a_2U + a_4U^2 + \cdots + a_{2k}U^k.$$

In other words, every even polynomial is a polynomial in X^2.

Now let us look at maps f: $\mathbf{R} \to \mathbf{R}$ that commute with the group G, that is, maps f satisfying

$$f(-X) = -f(X).$$

We call such maps *odd* functions. An argument similar to the one for even functions shows that odd polynomials f have the form

$$f(X) = b_1X + b_3X^3 + \cdots + b_{2k+1}X^{2k+1}.$$

It follows that if f is odd we may write

$$f(X) = (b_1 + b_3U + \cdots + b_{2k}U^k)X,$$

where, as above, $U = X^2$. Simply put, every odd polynomial can be written as the product of an even polynomial with X.

This decomposition generalizes to all compact symmetry groups of linear transformations of $\mathbf{R}^m$. The general observations are as follows.

1. Let $p : \mathbf{R}^m \to \mathbf{R}$ be G-invariant and let $f : \mathbf{R}^m \to \mathbf{R}^m$ be G-equivariant. Then $pf : \mathbf{R}^m \to \mathbf{R}^m$ is G-equivariant.
2. There exist a finite number of G-invariant polynomials
$$U_1(X),\ U_2(X),\ \ldots,\ U_s(X)$$
such that every G-invariant polynomial is a polynomial in $U_1(X), \ldots, U_s(X)$.
3. There exist a finite number of G-equivariant polynomial mappings
$$f_1(X), f_2(X), \ldots, f_t(X)$$
such that every G-equivariant polynomial mapping has the form
$$p_1(X) f_1(X) + \cdots + p_t(X) f_t(X)$$
where each $p_j(X)$ is a G-invariant polynomial.

In terms of even and odd polynomials, these statements translate as follows.

1. Even polynomials times odd polynomials are odd polynomials.
2. Every even polynomial is a polynomial in $U_1(X) = X^2$.
3. Every odd polynomial is an even polynomial times $f_1(X) = X$.

$\mathbf{D}_n$ *equivariants*

We recall that the action of $\mathbf{D}_n$, the planar symmetry group of the regular n-sided polygon, is generated by
$$\kappa z = \bar{z} \quad \text{and} \quad \rho z = e^{2\pi i/n} z,$$
where $z \in \mathbf{C}$ is a complex number.

Define the polynomial maps $U_1, U_2 : \mathbf{C} \to \mathbf{R}$ by
$$U_1(z) = z\bar{z} \quad \text{and} \quad U_2(z) = z^n + \bar{z}^n.$$

It is easy to check that both U_1 and U_2 are $\mathbf{D}_n$ invariant polynomials. For future reference, we note that $U_2(z) = 2\ \mathrm{Re}(z^n)$. We claim that every $\mathbf{D}_n$ invariant polynomial mapping $p : \mathbf{C} \to \mathbf{R}$ may be written as a polynomial in U_1 and U_2. That is, there is a polynomial map $q : \mathbf{R}^2 \to \mathbf{R}$ such that $p = q(U_1, U_2)$. We shall also show that if $f : \mathbf{C} \to \mathbf{C}$ is $\mathbf{D}_n$ equivariant then we may write
$$f = p_1 z + p_2\, \bar{z}^{n-1},$$
where p_1 and p_2 are $\mathbf{D}_n$ invariant polynomials. It is worth remarking, and useful for applications, that these expressions for $\mathbf{D}_n$ invariants and equivariants also hold for smooth (that is, infinitely differentiable) mappings with the change that q will now be a smooth map from $\mathbf{R}^2$ to $\mathbf{R}$ and p_1 and p_2 will be smooth invariant functions.

We start by verifying our claim about polynomial $\mathbf{D}_n$ invariants. Let $p : \mathbf{C} \to \mathbf{R}$ be a $\mathbf{D}_n$ invariant polynomial map. We may write p in the form
$$p(z) = \sum_{j,k \geq 0} a_{j,k} z^j \bar{z}^k,$$
where each $a_{j,k} \in \mathbf{C}$ is a complex number. Note that since p is a polynomial this sum is actually finite. The fact that p is real-valued ($p(z) = \overline{p(z)}$) implies $\overline{a_{j,k}} = a_{k,j}$. Since $p(\kappa z) = p(z)$ we also have $a_{j,k} = a_{k,j}$. It follows that all of the $a_{j,k}$ are real. Taking out multiples of $z\bar{z}$, we can write
$$p(z) = \sum_k b_k(z\bar{z})(z^k + \bar{z}^k),$$
where b_k is a polynomial in $z\bar{z}$. We now use the fact that $p(\rho z) = p(z)$ to conclude that $b_k = 0$ unless k is divisible by n. Thus p actually has the form
$$p(z) = \sum_k c_k(z\bar{z})(z^{nk} + \bar{z}^{nk}).$$

Finally, we use the identity
$$z^{n(l+1)} + \bar{z}^{n(l+1)} = (z^{nl} + \bar{z}^{nl})(z^n + \bar{z}^n) - (z\bar{z})^n (z^{n(l-1)} + \bar{z}^{n(l-1)})$$
and induction to observe that $p(z)$ can be written as a polynomial in U_1 and U_2.

Now suppose that $f : \mathbf{C} \to \mathbf{C}$ is a polynomial mapping that commutes with $\mathbf{D}_n$. Using complex notation, we can again write
$$f(z) = \sum_{j,k \geq 0} a_{j,k} z^j \bar{z}^k,$$
where each $a_{j,k} \in \mathbf{C}$. Here, though, we cannot assume that f is real-valued.

The fact that f commutes with κ means that $f(\bar{z}) = \overline{f(z)}$, from which it follows that $a_{j,k} = \overline{a_{j,k}}$. Hence all of the $a_{j,k}$ are real.

Since f commutes with ρ we have the identity

$$f(z) = e^{-2\pi i/n} f(e^{2\pi i/n} z).$$

It now follows that $a_{j,k} = 0$ unless $j \equiv k + 1 (\mathrm{mod}\ n)$. Grouping terms, we can rewrite f in the form

$$f(z) = \sum_{k \geq 0} [b_k(z\bar{z}) z^{nk+1} + c_k(z\bar{z}) \bar{z}^{n(k+1)-1}].$$

Applying the identities

$$z^{n(l+1)+1} = (z^{nl} + \bar{z}^{nl}) z - (z\bar{z}) \bar{z}^{nl-1},$$
$$\bar{z}^{n(l+1)-1} = (z^{nl} + \bar{z}^{nl}) \bar{z}^{n-1} - (z\bar{z})^{n-1} z^{n(l-1)+1},$$

and using induction, allows us to put f in the desired form.

Suppose that $f: \mathbf{C} \to \mathbf{C}$ is a $\mathbf{D}_n$ equivariant polynomial map. Truncating f to lowest order in U_1 and U_2 and replacing U_2 by $V_1 = \frac{1}{2} U_2$ yields the mapping

$$\begin{aligned} F(z) &= (\lambda + \alpha z\bar{z} + \beta \,\mathrm{Re}(z^n)) z + \gamma \bar{z}^{n-1} \\ &= (\lambda + \alpha U_1 + \beta V_1) z + \gamma \bar{z}^{n-1}, \end{aligned}$$

where λ, α, β, and γ are real parameters. This is the mapping we have used to produce most of our symmetric icons. Of course, our choice of this particular $\mathbf{D}_n$ equivariant mapping is somewhat arbitrary. However, it yields the simplest class of $\mathbf{D}_n$ equivariant mappings that display chaotic behaviour and symmetry creation.

In our investigations of maps with $\mathbf{D}_n$ symmetry, we have sometimes added an extra non-polynomial term

$$f(z) = (\lambda + \alpha U_1 + \beta V_1) z + \gamma \bar{z}^{n-1} + \delta \,\mathrm{Re}([z/|z|]^{np}) z |z|.$$

Although the extra term is $\mathbf{D}_n$ equivariant, it does have a mild singularity at the origin. In particular, for non-zero δ, this term tends to dominate the iteration near $z = 0$.

$\mathbf{Z}_n$ *equivariants*

We shall now briefly describe the $\mathbf{Z}_n$ invariants and equivariants. We start by observing that

$$U_1 = z\bar{z}, \quad V_1 = \mathrm{Re}(z^n), \quad V_2 = \mathrm{Im}(z^n)$$

are $\mathbf{Z}_n$ equivariants. Just as in the case of $\mathbf{D}_n$ equivariants, we may show that every $\mathbf{Z}_n$ invariant is a polynomial in U_1, V_1, and V_2.

Turning to $\mathbf{Z}_n$ equivariants, we may show that every $\mathbf{Z}_n$ equivariant map $f: \mathbf{C} \to \mathbf{C}$ may be written

$$f = p_1(U_1, V_1)\, z + p_2(U_1, V_1)\, \bar{z}^{n-1},$$

where p_1 and p_2 are arbitrary complex-valued polynomials.

In our numerical explorations of $\mathbf{Z}_n$ symmetry, we have added the term $\omega i z$ to the truncation we used for the study of $\mathbf{D}_n$ symmetry, to yield

$$f(z) = (\lambda + \alpha z\bar{z} + \beta \,\mathrm{Re}(z^n) + \omega i) z + \gamma \bar{z}^{n-1},$$

where λ, α, β, γ, and ω are real parameters. If $\omega = 0$ then f is $\mathbf{D}_n$ equivariant, while if $\omega \neq 0$ then f is only $\mathbf{Z}_n$ equivariant.

Appendix D

PLANAR LATTICES

A planar lattice is a subset $\mathcal{L}$ of $\mathbf{R}^2$ consisting of all vectors of the form

$$n_1\boldsymbol{k}_1 + n_2\boldsymbol{k}_2,$$

where $\boldsymbol{k}_1$ and $\boldsymbol{k}_2$ are noncollinear vectors in the plane and n_1 and n_2 are integers. The set of vectors $\{\boldsymbol{k}_1,\boldsymbol{k}_2\}$ is called a *basis* for $\mathcal{L}$. In Figure D.1, we have drawn part of the lattice defined by taking $\boldsymbol{k}_1$ and $\boldsymbol{k}_2$ to be the unit vectors along the x- and y-axes, respectively. For obvious reasons, this lattice is called a *square lattice*.

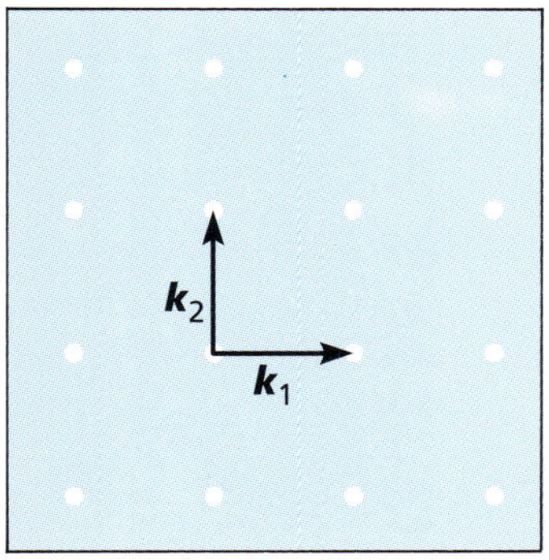

Figure D.1 *The square lattice.*

Suppose f is a mapping of the plane. We call f a *symmetry* of $\mathcal{L}$ if $f(\mathcal{L}) = \mathcal{L}$. Our first task is to describe the group of linear symmetries of $\mathcal{L}$.

Let A be a linear map of the plane. Since $\{\boldsymbol{k}_1, \boldsymbol{k}_2\}$ is a basis of $\mathbf{R}^2$, we can express $A(\boldsymbol{k}_1)$ and $A(\boldsymbol{k}_2)$ in terms of $\boldsymbol{k}_1$ and $\boldsymbol{k}_2$. Specifically, we may find real numbers a, b, c, and d such that

$$A(\boldsymbol{k}_1) = a\boldsymbol{k}_1 + b\boldsymbol{k}_2,$$
$$A(\boldsymbol{k}_2) = c\boldsymbol{k}_1 + d\boldsymbol{k}_2.$$

In order that A be a symmetry of $\mathcal{L}$, we must have $A(\boldsymbol{k}_1), A(\boldsymbol{k}_2) \in \mathcal{L}$. Hence, a, b, c, and d must be *integers*. Provided that a, b, c, and d are integers, it is easy to show that $A(n_1\boldsymbol{k}_1 + n_2\boldsymbol{k}_2) \in \mathcal{L}$ for all integers n_1 and n_2. However, for A to be a linear symmetry of $\mathcal{L}$, we also require $A(\mathcal{L})$ to be *equal* to $\mathcal{L}$. Clearly, it is not enough to require only that a, b, c, and d are integers (take $a = b = c = d = 0$). However, to be a symmetry, it suffices that the linear map A is invertible and that the inverse map of A is also a symmetry of $\mathcal{L}$. With a little more work, it follows that A is a linear symmetry of $\mathcal{L}$ if

(S1) a, b, c, and d are integers;
(S2) $ac - bd = \pm 1$.

We refer to the collection of linear maps satisfying conditions (S1) and (S2) as the group of linear symmetries of $\mathcal{L}$ and denote it by $G_{\mathcal{L}}$. In the case where the linear map A satisfies only (S1), we refer to A as a linear *endomorphism* of $\mathcal{L}$. We denote the space of linear endomorphisms $\mathcal{L}$ by $E_{\mathcal{L}}$. Obviously, $E_{\mathcal{L}} \supset G_{\mathcal{L}}$ and indeed $G_{\mathcal{L}}$ consists precisely of the invertible elements of $E_{\mathcal{L}}$.

Suppose that A is an element of $G_{\mathcal{L}}$. Then $\{A(\boldsymbol{k}_1), A(\boldsymbol{k}_2)\}$ is also a basis of $\mathcal{L}$. Since $G_{\mathcal{L}}$ is a rather large group, it follows that a lattice can have very many different bases. In Figure D.2, we show another basis for the square lattice found by taking $a = 2$ and $b = c = d = 1$.

Up to this point we have not introduced any geometry into our discussion of lattices. Indeed, it follows from (S1) and (S2) that if $\mathcal{L}$ and $\mathcal{L}'$ are any two lattices, then $G_{\mathcal{L}}$ and $G_{\mathcal{L}'}$ are isomorphic; and so, at least from the point of view of linear symmetries, all lattices have the same structure.

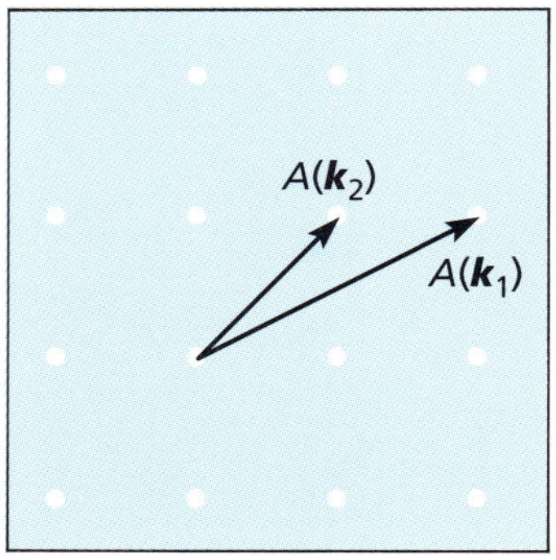

Figure D.2 An alternate basis for the square lattice.

The way we introduce geometric structure is to regard $\mathbf{R}^2$ as a Euclidean vector space. In particular, we shall assume that $\mathbf{R}^2$ comes with its standard inner product. If $x, y \in \mathbf{R}^2$, we let $x \cdot \mathbf{y}$ denote the *inner* (or *dot*) *product* of x with y. We let $\mathbf{O}(2)$ denote the orthogonal group consisting of the set of all linear transformations of $\mathbf{R}^2$ that preserve the inner product. Otherwise stated, $\mathbf{O}(2)$ is the group consisting of all rotations about the origin of $\mathbf{R}^2$ together with all reflections in lines through the origin.

We define the *holohedry* of the lattice $\mathcal{L}$ to be the group $\mathcal{H}_{\mathcal{L}}$ consisting of all those elements of $G_{\mathcal{L}}$ which are rotations and reflections; that is,

$$\mathcal{H}_{\mathcal{L}} = G_{\mathcal{L}} \cap \mathbf{O}(2).$$

Unlike what happened in our discussion of $G_{\mathcal{L}}$, the group $\mathcal{H}_{\mathcal{L}}$ depends on the particular lattice $\mathcal{L}$.

It is not hard to show that $\mathcal{H}_{\mathcal{L}}$ is always finite and does not depend on the choice of basis of $\mathcal{L}$. Moreover, since $-I$ is always a linear symmetry of $\mathcal{L}$, we see that $\mathcal{H}_{\mathcal{L}}$ always contains $\pm I$. With a bit more work, one can show that there are precisely *five* groups that can occur as the holohedry of a planar lattice. We describe these groups, and representative lattices, in Table D.1.

Table D1 The planar lattices

Name	Length of basis vectors	Angle between k_1 and k_2	Holohedry
Oblique	$\lvert k_1\rvert \neq \lvert k_2\rvert$	$\alpha \neq 90°$	$\mathbf{Z}_2$
Rectangular	$\lvert k_1\rvert \neq \lvert k_2\rvert$	$\alpha = 90°$	$\mathbf{Z}_2 \oplus \mathbf{Z}_2$
Rhombus	$\lvert k_1\rvert = \lvert k_2\rvert$	$\alpha \neq 60°, 90°$	$\mathbf{Z}_2 \oplus \mathbf{Z}_2$
Square	$\lvert k_1\rvert = \lvert k_2\rvert$	$\alpha = 90°$	$\mathbf{D}_4$
Hexagonal	$\lvert k_1\rvert = \lvert k_2\rvert$	$\alpha = 60°$	$\mathbf{D}_6$

Fundamental cells and tori

Suppose that the lattice $\mathcal{L}$ has basis $\{k_1, k_2\}$. The *fundamental cell* or *period parallelogram* of the lattice is the parallelogram spanned by k_1 and k_2. The plane is tessellated by translations of the fundamental cell using vectors in $\mathcal{L}$.

As we have already pointed out, a given lattice has many different bases. On the square and hexagonal lattices, however, certain of these bases better reflect the symmetry of the holohedry. In particular, this happens when the basis vectors have the same length and lie on axes of symmetry of the holohedry. Consequently, on the square lattice we choose basis vectors

$$k_1 = (1, 0), \qquad k_2 = (0, 1),$$

while on the hexagonal lattice we choose basis vectors

$$k_1 = (1, 0), \qquad k_2 = (\tfrac{1}{2}, \tfrac{\sqrt{3}}{2})$$

At this point, we note that the opposite sides of the fundamental cell are identified by translations. When we think of this process of identifying opposite sides geometrically, we produce a *torus* denoted by T^2. We refer the reader to Figure D.3 for the case of the square lattice.

Suppose we have a mapping $f\colon \mathbf{R}^2 \to \mathbf{R}^2$. When does f induce a mapping on the torus? For this to happen, the points $f(X)$ and $f(X+k)$, where $k \in \mathcal{L}$, must differ by only a lattice element; that is, for each $k \in \mathcal{L}$, there exists a $k \in \mathcal{L}$ such that

$$f(X + k) = f(X) + \hat{k}.$$

We can rephrase the definition of $E_{\mathcal{L}}$ in this language: the space $E_{\mathcal{L}}$ of linear endomorphisms of $\mathcal{L}$ consists of those linear maps of $\mathbf{R}^2$ that induce a mapping on the torus. To see this, let A be a linear map on $\mathbf{R}^2$. Then A induces a map on T^2 if and only if, for every $k \in \mathcal{L}$, the vector $A(X+k) - A(X) = A(k)$ is in $\mathcal{L}$.

Action of the holohedry on the torus

We assume that $\mathcal{L}$ is either the square or hexagonal lattice and take the associated fundamental cell as described above. Let T^2 denote the torus defined by the fundamental cell. Since

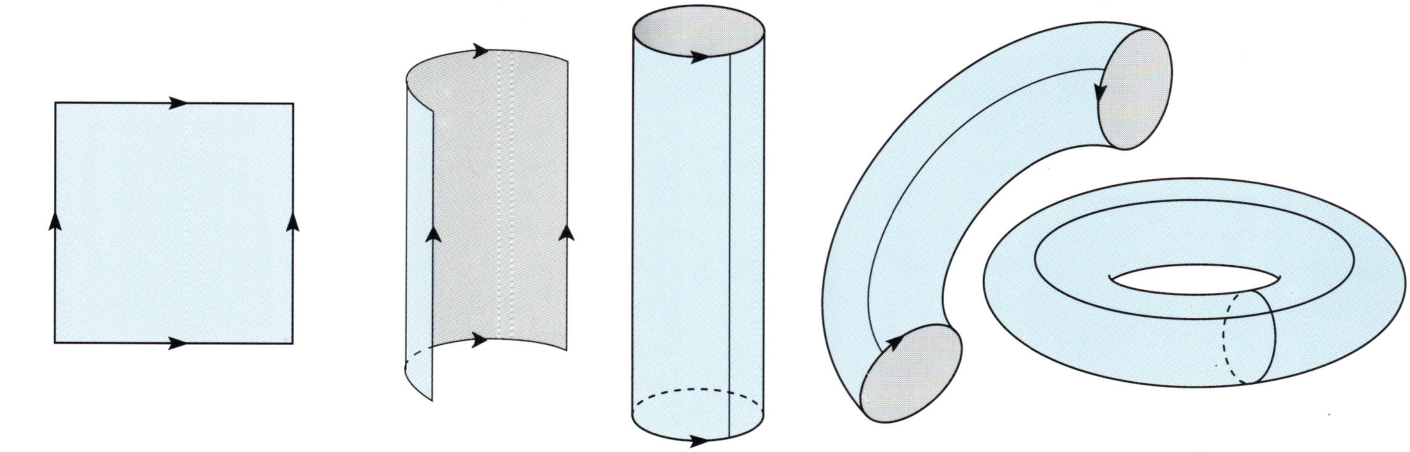

Figure D.3 *Identifying the opposite sides of a fundamental cell to make a torus.*

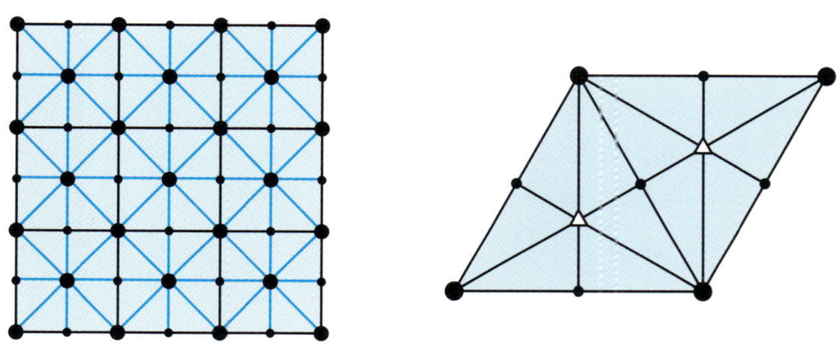

(*a*) (*b*)

Figure D.4 *Lines of symmetry on the fundamental cell (large dots are used to denote points fixed by the holohedy and small dots and triangles to denote points fixed by proper subgroups of the holohedry): (a) square lattice (b) hexagonal lattice.*

$\mathcal{H}_{\mathcal{L}} \subset E_{\mathcal{L}}$ it follows that $\mathcal{H}_{\mathcal{L}}$ acts on T^2. We wish to describe the lines of symmetry for the action of $\mathcal{H}_{\mathcal{L}}$ on T^2. This is most easily done by drawing the lines of symmetry on the fundamental cell. We have done this for the fundamental cells of the square and hexagonal lattices in Figure D.4.

The reader should note that there are two points on the fundamental cell of the square lattice that are fixed by all elements of the holohedry. These points are (0, 0) and $(\frac{1}{2}, \frac{1}{2})$; they are characterized as being the only points that lie on all the axes of symmetry of the action of $\mathcal{H}_{\mathcal{L}}$. On the other hand, if we take the hexagonal lattice, then the action of $\mathcal{H}_{\mathcal{L}}$ on T^2 has just one fixed point: (0, 0).

We gain more insight into the symmetry structure of the action of the holohedry by drawing the associated tessellations, with lines of symmetry. We show this in Figure D.5(a) for the

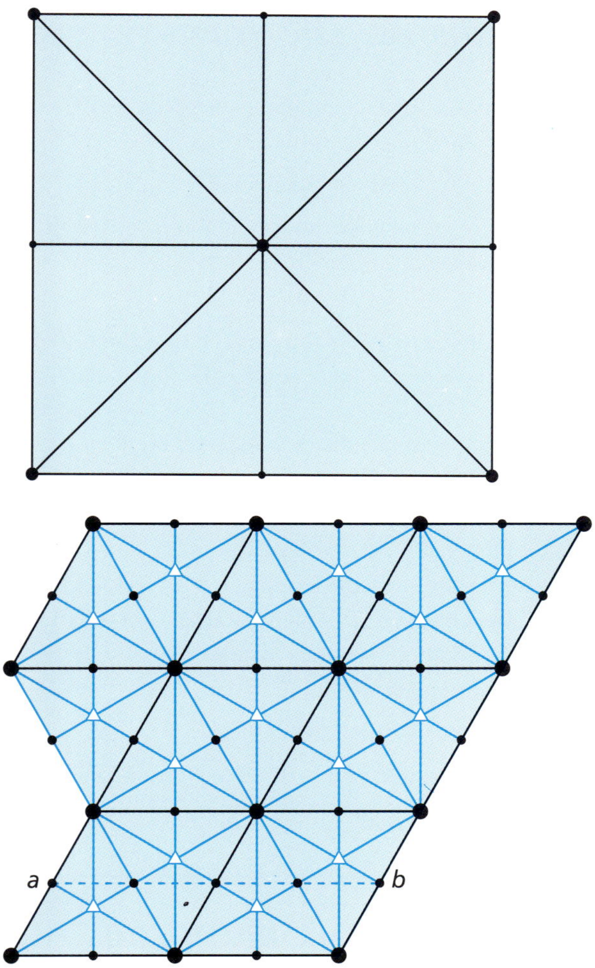

(*a*) (*b*)

Figure D.5 *Lines of symmetry in the plane (see Figure D.4 for an explanation of notation): (a) square lattice (b) hexagonal lattice.*

square lattice and in Figure D.5(b) for the hexagonal lattice. The reader should note that there is much more structure than appears in the corresponding pictures for the linear actions of the holohedry groups $\mathbf{D}_4$ and $\mathbf{D}_6$ on the plane. The lines of symmetry that occur in the tessellations are, of course, the lines of symmetry for the associated lattice. In the case of the hexagonal lattice, we see not only reflectional and rotational symmetries but also glide reflection symmetries. For example, referring to Figure D.5(b), we see there is a glide reflection symmetry defined by translation and reflection in the line *ab*. It is this additional structure that is, in part, responsible for the richness of the patterns seen in our pictures of chaotic quilt tilings.

Mappings on the torus

In general, every continuous mapping $f: \mathbf{R}^2 \to \mathbf{R}^2$ that induces a map on T^2 has the form

$$f(\boldsymbol{X}) = q(\boldsymbol{X}) + A(\boldsymbol{X}) + \boldsymbol{v}, \qquad \text{(D.1)}$$

where A is in $E_{\mathcal{L}}$, $\boldsymbol{v}$ is in the fundamental cell, $q(\mathbf{0}) = \mathbf{0}$, and q is $\mathcal{L}$-periodic; that is,

$$q(\boldsymbol{X} + \boldsymbol{k}) = q(\boldsymbol{X})$$

for all $\boldsymbol{k} \in \mathcal{L}$. We now verify formula (D.1) in three steps.

First note that $\boldsymbol{v}$ must equal $f(\mathbf{0})$; just evaluate (D.1) at $\boldsymbol{X} = \mathbf{0}$.

Second, since f induces a mapping on T^2 it follows that $f(\boldsymbol{k}_1) = \hat{\boldsymbol{k}}_1$ and $f(\boldsymbol{k}_2) = \hat{\boldsymbol{k}}_2$, where $\hat{\boldsymbol{k}}_1, \hat{\boldsymbol{k}}_2 \in \mathcal{L}$. Define A to be the linear map satisfying $A(\boldsymbol{k}_j) = \hat{\boldsymbol{k}}_j$ for $j = 1, 2$. By construction A lies in $E_{\mathcal{L}}$.

Third, note that the mappings $f(\boldsymbol{X})$ and $f(\boldsymbol{X}) + \boldsymbol{k}$ induce the same mappings on T^2. Hence, after a translation by an element of the lattice, we may assume that $\boldsymbol{v}$ is in the fundamental cell. Finally, we define

$$q(\boldsymbol{X}) \equiv f(\boldsymbol{X}) - A(\boldsymbol{X}) - \boldsymbol{v}.$$

By construction $q(\mathbf{0}) = \mathbf{0}$. Since $f(\boldsymbol{X} + \boldsymbol{k}_j) = f(\boldsymbol{X}) + \hat{\boldsymbol{k}}_j$, it follows that $q(\boldsymbol{X} + \boldsymbol{k}_j) - q(\boldsymbol{X}) = \mathbf{0}$. So q is $\mathcal{L}$-periodic as claimed.

Symmetric torus mappings

We assume that the planar mapping f induces a mapping on the torus and hence has the form (D.1). As noted above, the holohedry $\mathcal{H}_{\mathcal{L}}$ of the lattice acts on the torus; we ask which mappings of the torus will commute with the action of the holohedry on the torus. These are the mappings that will have attractors with square symmetry in the fundamental cell of the square lattice and hexagonal symmetry on the fundamental cell of the hexagonal lattice; indeed, these are the mappings that will produce the quilt patterns we have shown in previous chapters.

For f to commute with the action of the holohedry on the torus, it must commute modulo the lattice with the action of $\mathcal{H}_{\mathcal{L}}$ on the plane; that is,

$$f(h\boldsymbol{X}) = hf(\boldsymbol{X}) + \boldsymbol{k}_h,$$

where $h \in \mathcal{H}_{\mathcal{L}}$ and $\boldsymbol{k}_h \in \mathcal{L}$. On the square and hexagonal lattices this commutativity takes place precisely when f has the form

$$f(\boldsymbol{X}) = q(\boldsymbol{X}) + m\boldsymbol{X} + \boldsymbol{v}, \qquad \text{(D.2)}$$

where m is an integer, $\boldsymbol{v}$ is a fixed point of the action of $\mathcal{H}_{\mathcal{L}}$ when projected onto the torus, $q(\mathbf{0}) = \mathbf{0}$, and q is an $\mathcal{L}$-periodic mapping that commutes with $\mathcal{H}_{\mathcal{L}}$. To verify (D.2), use the form (D.1) to compute $f(h\boldsymbol{X}) - hf(\boldsymbol{X}) = \mathbf{0}$ and obtain

$$q(h\boldsymbol{X}) - hq(\boldsymbol{X}) = hA(\boldsymbol{X}) - A(h\boldsymbol{X}) + h\boldsymbol{v} - \boldsymbol{v} + \boldsymbol{k}_h.$$

Observe that the left-hand side of this equation is $\mathcal{L}$-periodic (using the fact that h is in the holohedry) and the right-hand side is affine and unbounded unless the linear part is identically zero. Hence,

$$hA(\boldsymbol{X}) = A(h\boldsymbol{X}).$$

A short calculation shows that the only linear mappings on the plane that commute with either $\mathbf{D}_4$ or $\mathbf{D}_6$ are multiples of the identity. For such mappings to be in the holohedry, they must be integer multiples of the identity. Hence $A(\boldsymbol{X}) = m\boldsymbol{X}$ as claimed.

Next we use the fact that $q(\mathbf{0}) = \mathbf{0}$ to see that $h\boldsymbol{v} = \boldsymbol{v} - \boldsymbol{k}_h$ for all elements of the holohedry. Hence $h\boldsymbol{v} = \boldsymbol{v}$ when projected onto the torus; that is, $\boldsymbol{v}$ is a fixed point of the holohedry on the torus, as claimed. It now follows that q commutes with the action of the holohedry on the plane.

As we have already shown, $\boldsymbol{v}$ is $\mathbf{0}$ on the hexagonal lattice and either $\mathbf{0}$ or $(\frac{1}{2}, \frac{1}{2})$ on the square lattice.

In order to specify the family of mappings that may produce the types of periodic quilts that we desire, we must enumerate the $\mathcal{L}$-periodic mappings that commute with the holohedry. This is most easily done using Fourier series.

Fourier expansions of $\mathcal{L}$-periodic mappings

There is a simple but ingenious way of enumerating $\mathcal{L}$-periodic functions using what is called the *dual lattice*. The idea is to ask which *plane waves* $\exp(2\pi i \boldsymbol{l} \cdot \boldsymbol{X})$ are $\mathcal{L}$-periodic mappings. A quick calculation shows that periodicity holds precisely when $\boldsymbol{l} \cdot \boldsymbol{k}$ is an integer for all lattice vectors $\boldsymbol{k} \in \mathcal{L}$. Indeed, the set of all such $\boldsymbol{l}$ forms a lattice called the *dual lattice* and is denoted by $\mathcal{L}^\star$.

A basis for the dual lattice $\{\boldsymbol{l}_1, \boldsymbol{l}_2\}$ satisfies

$$\boldsymbol{l}_i \cdot \boldsymbol{l}_j = \delta_{i,j},$$

where $\delta_{i,j}$ is the usual *Kronecker* delta, which equals 1 if $i = j$, and equals 0 otherwise. For the square lattice (with $\boldsymbol{k}_1 = (1, 0)$ and $\boldsymbol{k}_2 = (0, 1)$), we have

$$\boldsymbol{l}_1=(1, 0) \quad \text{and} \quad \boldsymbol{l}_2=(0, 1).$$

For the hexagonal lattice (with $\boldsymbol{k}_1 = (1, 0)$ and $\boldsymbol{k}_2 = (-\frac{1}{2}, \frac{\sqrt{3}}{2})$), we have

$$\boldsymbol{l}_1=(1, -\tfrac{1}{\sqrt{3}}) \quad \text{and} \quad \boldsymbol{l}_2=(0, -\tfrac{2}{\sqrt{3}}).$$

We can now write down a large number of $\mathcal{L}$-periodic mappings of the plane to the plane. For each $\boldsymbol{l} \in \mathcal{L}^\star$, let $z_{\boldsymbol{l}}$ be in $\mathbf{C}^2$. Then the formal sum

$$q(\boldsymbol{X}) = \sum_{\boldsymbol{l} \in \mathcal{L}^\star} \mathrm{Re}[\exp(2\pi i \boldsymbol{l} \cdot \boldsymbol{X}) z_{\boldsymbol{l}}]$$

is $\mathcal{L}$-periodic, when the sum converges.

It follows from Fourier analysis, that every continuous q can be written in this form. We won't go into the details here, but rather ask what restrictions have to be put on the *amplitude* vectors $z_{\boldsymbol{l}}$ for q to commute with the holohedry.

Indeed, for q to commute with the holohedry, we must have

$$hq(h^{-1}\boldsymbol{X}) = q(\boldsymbol{X})$$

for all $h \in \mathcal{H}_{\mathcal{L}}$. We now compute formally

$$hq(h^{-1}\boldsymbol{X}) = \sum_{\boldsymbol{l} \in \mathcal{L}^\star} \mathrm{Re}[\exp(2\pi i \boldsymbol{l} \cdot h^{-1}\boldsymbol{X})\, hz_{\boldsymbol{l}}]$$

$$= \sum_{\boldsymbol{l} \in \mathcal{L}^\star} \mathrm{Re}[\exp(2\pi i h\boldsymbol{l} \cdot \boldsymbol{X})\, hz_{\boldsymbol{l}}]$$

Equating terms in the Fourier expansion of $q(\boldsymbol{X})$ and $hq(h^{-1}\boldsymbol{X})$ yields

$$hz_{\boldsymbol{l}} = z_{h\boldsymbol{l}}.$$

Next we observe that, since $-I$ is in the holohedry,

$$z_{-\boldsymbol{l}} = -z_{\boldsymbol{l}}.$$

It follows by combining the $\boldsymbol{l}$ and $-\boldsymbol{l}$ terms that we can write q as

$$q(\boldsymbol{X}) = \sum_{\boldsymbol{l} \in \mathcal{L}^\star} \sin(2\pi \boldsymbol{l} \cdot \boldsymbol{X})\, \boldsymbol{a}_{\boldsymbol{l}},$$

where $\boldsymbol{a}_{\boldsymbol{l}} \in \mathbf{R}^2$ and

$$\boldsymbol{a}_{h\boldsymbol{l}} = h\boldsymbol{a}_{\boldsymbol{l}} \tag{D.3}$$

for all $\boldsymbol{l} \in \mathcal{L}^\star$.

We can now rewrite the summation that forms q in a slightly more efficient way. The holohedry $\mathcal{H}_{\mathcal{L}}$ acts on the dual lattice $\mathcal{L}^\star$. Thus we can think of the dual lattice as being partitioned by the orbits of the action of the holohedry. Suppose we form a set of dual lattice vectors $\bar{\mathcal{L}}^\star$ that consists of one dual lattice vector from each group orbit in $\mathcal{L}^\star$. Then we can write

$$q(\boldsymbol{X}) = \sum_{\boldsymbol{l} \in \bar{\mathcal{L}}^\star} \left(\sum_{h \in \mathcal{H}_{\mathcal{L}}} \sin(2\pi \boldsymbol{l} \cdot \boldsymbol{X})\, h\boldsymbol{a}_{\boldsymbol{l}} \right).$$

Of course, since $\boldsymbol{a}_{-\boldsymbol{l}} = -\boldsymbol{a}_{\boldsymbol{l}}$ and $\sin(-x) = -\sin(x)$ we need only write the inner sum over half of the elements of the holohedry. We end by writing out explicitly a few terms in this Fourier expansion of q for both the hexagonal and square lattices.

The quilt mappings

We begin with the square lattice. On the square lattice, the dual basis is simply $\boldsymbol{l}_1 = \boldsymbol{k}_1$, $\boldsymbol{l}_2 = \boldsymbol{k}_2$. The first group orbit of wave vectors is just $\pm\boldsymbol{l}_1$, $\pm\boldsymbol{l}_2$, as pictured in Figure D.6(a). Since there is a reflection in the holohedry that leaves each of these vectors invariant, we get the term

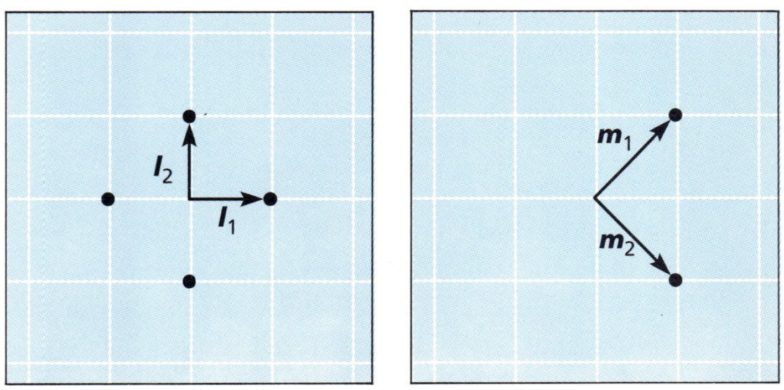

(*a*) (*b*)

Figure D.6 Group orbits of dual wave vectors on the square lattice.

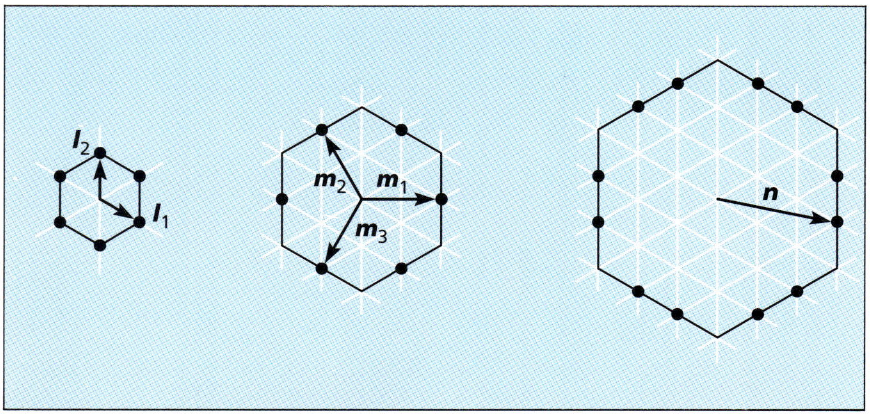

(*a*) (*b*) (*c*)

Figure D.7 *Group orbits of dual wave vectors on the hexagonal lattice.*

$$\lambda(\sin(2\pi\, l_1 \cdot X)l_1 + \sin(2\pi\, l_2 \cdot X)l_2).$$

If we write $X = (x, y)$ in coordinates, then this term is just

$$\lambda(\sin 2\pi x, \sin 2\pi y).$$

For the second term, we choose the orbit generated by the wave vector $m_1 = l_1 + l_2$ (see Figure D.6(b)). The group orbit of this vector is m_1 and $m_2 = l_2 - l_1$ and their negatives, and the associated term in q is

$$\alpha[\sin(2\pi\, m_1 . X)m_1 + \sin(2\pi\, m_2 . X)m_2].$$

If one unravels all of the vector notation, one arrives at the term

$$\alpha(\sin 2\pi x \cos 2\pi y, \sin 2\pi y \cos 2\pi x).$$

Continuing along these lines, we may show that the general formula for $q(x, y)$ in coordinates is

$$q(x, y) = (p(x, y), p(y, x)),$$

where

$$p(x, y) = \sum_{m \geqslant 0,\, n > 0} a_{m,n} \sin 2\pi mx \cos 2\pi ny.$$

Arbitrarily, we chose four terms from the Fourier series for p:

$$p(x, y) = \lambda \sin 2\pi x + \alpha \sin 2\pi x \cos 2\pi y + \beta \sin 4\pi x + \gamma \sin 6\pi x \cos 4\pi y.$$

For our investigations of square quilts, we take this mapping together with the translation term v and the linear term mX.

We conclude with the hexagonal lattice. Choose the vector l_1 to be an orbit representative and let $l_3 = -l_1 - l_2$. Then the first term in our expansion of q has the form

$$\sin(2\pi l_1 \cdot X)a_1 + \sin(2\pi l_2 \cdot X)a_2 + \sin(2\pi l_3 \cdot X)a_3. \quad (D.4)$$

As shown in Figure D.7(a), there is an element h of the holohedry that fixes l_2 and interchanges l_1 with l_3. It follows that

$$ha_{l_2} = a_{hl_2} = a_{l_2}.$$

Since the only vectors fixed by h are scalar multiples of l_2, it follows that $a_2 = \lambda l_2$ for some real number λ. Hence we may rewrite (D.4) as

$$\lambda[\sin(2\pi l_1 \cdot X)l_1 + \sin(2\pi l_2 \cdot X)l_2 + \sin(2\pi l_3 \cdot X)l_3].$$

Similarly, if we let $m_1 = 2l_1 + l_2$, we obtain a group orbit as shown in Figure D.7(b). Note that there is a reflectional symmetry in the holohedry that fixes m_1, so that the same argument just given applies. Hence, the second term for q that we write down is

$$\alpha[\sin(2\pi\, m_1 \cdot X)m_1 + \sin(2\pi\, m_2 \cdot X)m_2 + \sin(2\pi\, m_3 \cdot X)m_3],$$

where m_1, m_2, and m_3 are as shown in Figure D.7(b).

For the third (and last) term we take $n = 3l_1 + 2l_2$. None of the nontrivial elements of the holohedry fixes this dual wave vector. Hence we get a total of six sine terms. Let R denote rotation counterclockwise by 120° and let F denote reflection in the x-axis. Then the new term is

$$\sin(2\pi n \cdot X)a_n + \sin(2\pi Rn \cdot X)Ra_n + \sin(2\pi R^2 n \cdot X)R^2 a_n + \sin(2\pi Fn \cdot X)Fa_n + \sin(2\pi RFn \cdot X)RFa_n + \sin(2\pi R^2 Fn \cdot X)R^2 Fa_n,$$

where $a_n = (\beta, \gamma)$ is now an arbitrary two-vector.

Figure D.8 (opposite) *Hex Nuts.*

Summing these various terms together gives a mapping with four real parameters (λ, α, β, γ) and one integer m. Together with the term $m\mathbf{X}$, this is the mapping that we have used to produce the pictures of hexagonal quilts.

Quilts with cyclic symmetry

Several of the quilts we show have cyclic rather than dihedral symmetry. To obtain these quilts, we allow the parameters to take complex rather than real values. For example, in our investigations of square quilts with cyclic symmetry, we take the initial term to be

$$\lambda\,(\sin 2\pi x, \sin 2\pi y) + \omega\,(-\sin 2\pi y, \sin 2\pi x)$$

and allow ω to take nonzero values. We make a similar change in the case of hexagonal quilts.

There is a second way in which cyclic symmetry can be forced in the quilt mappings. The linear map A used in the derivation of our formulas was forced by dihedral symmetry to be an integer multiple of the identity. Cyclic symmetry only forces it to be an integer multiple of a root of unity. In particular, cyclic symmetry will appear in square quilts if the parameter m is chosen to be an integer multiple of i. Cyclic symmetry will appear in hexagonal quilts if m is chosen to be an integer multiple of $e^{i\pi/3}$.

We end this appendix with two symmetric chaos quilts based on a hexagonal lattice. The first quilt (Figure D.8) has $\mathbf{Z}_6$ symmetry and uses for the first time a non-integer value of the parameter m. The second quilt (Figure D.9) also has $\mathbf{Z}_6$ symmetry, but is computed using a nonzero value for ω.

Figure D.9 (opposite) *Marching Troupe.*

Bibliography

Benno Artmann, The Cloisters of Hauterive, *The Mathematical Intelligence* **13** No. 2 (1991) 44–49.

Julian Barnard, *Victorian Ceramic Tiles*, Studio Vista, London, 1972.

Michael Barnsley, *Fractals Everywhere*, Academic Press, San Diego, 1988.

Roger Bell, *Great Marques: Mercedes-Benz*, Octopus Books Ltd, London, 1980.

F. Bianchini and F. Corbetta, *The Kindly Fruits*, (English Edition) Cassell, London, 1977.

Rafael Bombelli da Bologna, *L'Algebra*, Prima edizione integrale, Feltrinelli, Milano, 1966.

Anke Brandstater and Harry L. Swinney, *Phys. Rev. A* **35** (1987) 2207–2222.

G. Cardano, *The Great Art or the Rules of Algebra*, tr. T. R. Wilmer, Cambridge, Mass., 1968.

Pascal Chossat and Martin Golubitsky, Symmetry increasing bifurcations of chaotic attractors, *Physica D* **32** (1988) 423– 436.

Painton Cowen, *Rose Windows*, Thames & Hudson, Ltd, London, 1979

John Crossley, *The Emergence of Number*, Upside Down A Book Company, Box 226, Yarra Glen, Victoria, Australia 3168, 1980.

Michael Dellnitz, Martin Golubitsky and Ian Melbourne, Mechanisms of symmetry creation. In: *Bifurcation and Symmetry* (E. Allgower, K. Bohmer and M. Golubitsky, eds.), ISNM **104**, Birkhäusser, Basel, 1992, 99–109.

Walter B. Denny, *The Ceramics of the Mosque of Rüstem Pasha and the Environment of Change*, Garland Publishers, Inc., New York, 1977.

Maurits C. Escher, *The Graphic Work of M. C. Escher*, Hawthorn/ Ballantine, New York, 1971.

Mitchell Feigenbaumm, Quantitative universality for a class of nonlinear transformations, *J. Stat. Phys.* **19** (1978) 25–52.

James Gleich, *Chaos: Making a New Science*, Viking Penguin Inc., New York, 1987.

Ernst J. Grube, *Islamic Pottery*, Faber & Faber, Ltd, London, 1976.

A. N. J. Th. a Th. van der Hoop, *Indonesian Ornamental Design*, A. C. Nix & Co., Bandoeng, 1949.

Claude Humbert, *Islamic Ornamental Design*, Faber & Faber, Ltd, London, 1980.

Glenn James and Robert C. James, *Mathematics Dictionary*, D. Van Nostrand Company, Inc., Princeton, 1960.

Carsten Knudsen, Rasmus Feldberg and Hans True, Bifurcations and chaos in a model of a rolling railway wheelset, *Phil. Trans. R. Soc. Lond. A* (1992). To appear.

Benoit Mandelbrot, *The Fractal Geometry of Nature*, W.H. Freeman & Co., San Francisco, 1982.

Karl Maramorosch (ed.), *The Atlas of Insect and Plant Viruses: Vol. 8*, Ultra Structures in Biological Systems, 1977.

Robert May, Simple mathematical models with very complicated dynamics, *Nature* **261** (1976) 459–467.

Franz Sales Meyer, *A Handbook of Ornament*, Duckworth & Co. Ltd, London, 1974.

Qi Ouyang and Harry L. Swinney, Transitions to chemical turbulence. *Nature* **352** (1992) 610–612.

Heinz-Otto Peitgen and Peter H. Richter, *The Beauty of Fractals, Springer-Verlag, Berlin, 1986.*

Ian Stewart, *Does God Play Dice?,* Basil Blackwell, London, 1990.

Ian Stewart and Martin Golubitsky, *Fearful Symmetry: Is God a Geometer?*, Basil Blackwell, London, 1992.

J. Triska, *The Hamlyn Encyclopedia of Plants*, Hamlyn, 1975.

Dorothy K. Washburn and Donald W. Crowe, *Symmetries of Culture*, University of Washington Press, Seattle, 1988.

André Weil, *Number Theory*, *An Approach Through History,* Birkhäuser, Boston, Basel, Stuttgart, 1984.

Hermann Weyl, *Symmetry*, Princeton University Press, Princeton, New Jersey, 1952.

Alfred North Whitehead, *Science and the Modern World,* Macmillan, New York, 1954.

Index

Figure Acknowledgements

Wendy Aldwyn **1.3, 1.4, 1.5, 1.6, 1.10, 1.11, 1.12, 1.14, 1.19, 1.20, 2.1, 2.2, 2.4, 2.7, 2.8, 2.10, 2.11, 3.3, 5.6, 5.7, 5.8, 5.9, 5.10, 5.11, 7.5, 7.6, 7.11, D.2, D.3, D.4, D.5, D.6, D.7**
Science Photo Library **1.9** (snowflake), **3.1** (Eric Grave), **3.8(a)** (Claude Nuridsany and Marie Perennon), **3.8(b)** (Manfred Kage)
Harry L. Swinny and Randy Tagg **1.21**
Harry L. Swinny and Anke Brandstater **1.22**
Cordon Art **2.14** (Symmetry drawing A by Maurits Escher)
Ancient Art and Architecture Collection **2.15**, **3.18(a)** (Chris Hellier), **3.27(b)**
Oxford Scientific Film **3.2(a)** (Geoff Kidd), **3.2(b)** (C. W. Helliwell)
Benno Artmann **3.4(a)**
United States Air Force photograph by Msgt. Ken Hammond **3.6(a)**
F. Bianchini and F. Corbetta, The Kindly Fruits, (English Edition, Cassell, London 1977 **3.7(a)**
Telegraph Colour Library **3.9(a)**
Christies, London **3.10(b)**, **3.10(d)**
Ernest J. Grube, Islamic Pottery, Faber & Faber Ltd, London 1980 **3.10(a)**
Painton Cowen, Rose Windows, Thames & Hudson Ltd, London 1979 **3.12**, **3.13(a)**
Karl Maramorosch (Ed.), T*he Atlas of Insect and Plant Viruses: Vol. 8, Ultra Structures in Biological Systems*, 1977 **3.16(a, b)**
Qi Ouyang and Harry L. Swinney **3.17**
The Gazebo of New York, Inc. **3.19**
A. N. J. Th. a Th. van der Hoop, *Indonesian Ornamental Design*, A.C. Nix & Co., Bandoeng, 1949 **3.20(a)**
Royal Doulton **3.23(a, b)**
Julian Barnard, *Victorian Ceramic Tiles*, Studio Vista, London, 1972 **3.23(c, d)**
Franz Sales Meyer, *A Handbook of Ornament*, Ducworth & Co. Ltd., London, 1974 **3.22**
Woodmansterne **3.25(a)**
Owen Jones, *Grammar of Ornament*, Dover, New York **3.26**
Michael Holford, Buildings Spain/Cordoba **3.27(a)**
British Museum **3.28(a)**
Sidney Harris **7.2**